Homebuilt

Toys

by

Gary F. Hartman

Published by
Gary F. Hartman
Lebanon, OR

Homebuilt Toys

Copyright © 2015

All rights reserved. No part of this
book may be reproduced or transmitted
in any form, or by any means
(except for the inclusion of brief
quotations in reviews) without the prior
written permission of the
author.

ISBN 978-0-9815399-5-9
Library of Congress
Control Number: 2015901659

Printed in U.S.A. By Lightning Source Inc.

Foreword to parents

In 2008, I wrote "Kids' Book of Adventure Projects" with the idea of getting kids interested in doing toys of their own from scraps of wood, paper, objects found around the house; essentially things most families pay no attention to, waste items that are usually tossed away into the garbage or disposed of during clean ups.

That book was all black and white with drawings, but this book is well illustrated with photos as well, plus this book is a larger format and in color.

As a kid, my parents supported my endless reading and endless projects, I made toys from anything and everything. They provided me with a Coping Saw and pocketknife early in my life and those tools gave me an endless supply of projects, probably soap carving at the beginning, then small wooden toys.

We moved often as my Stepfather was in construction work, and in every town there was a wonderful library providing an endless supply of project ideas and knowledge. No kid has an excuse not to learn, good schools or bad, THERE IS ALWAYS A LIBRARY.

We must encourage kids at an early age to learn, to search out and answer questions through books. The source is there for the asking... everywhere. There is no excuse for poor education.

It is sad that nowadays kids have been steered into television, video games, and provided with cell phones at an early age.

Back in the 1950's when I grew up, there was no such entertainment in every single home; kids played baseball amongst themselves, (no Little League... parents did not butt into the kids' activities...) the kids built toys, and tree houses themselves, they played "kick the can"; they developed imagination and ingenuity towards finding things to do-- they learned by doing--- without parents planning every activity in their lives.

Now, you often see kids traipsing down the street with their nose in an I-pad, cell phone or some device, not even watching the traffic or where they are going.

In the 1950's a kid would have given his favorite toy to have a walkie-talkie … I remember seeing ads in comic books, but of course those were string-in-between toy units; it was into the early 60's before real walkie-talkies became available from Allied Radio as a Knight Kit. Those were the days.

But desire meant a kid developed his imagination towards all manner of making toys or things to play with because parents couldn't afford buying them.

I hope a few kids will have the thrill of making their own toys from ideas presented here... I hope they will cherish and value the satisfaction, skill and confidence gained from these few ideas. As they mature, those ideas and skills will carry them into adulthood.

Gary F. Hartman
Lebanon, OR
www.jgenasplace.com

Previous Books:

Kids' Book of Adventure Projects	2008
Homebuilt Firearms	2010
Homebuilt Clocks	2011
Homebuilt .45 ACP Carbine	2013

Table of Contents

Chapter			
1	Starting Out		1
2	A Crawler Toy		10
3	Two Army Tanks		14
4	Historic Model Airplane		37
5	Pull Toy Dogs		75
6	Building a Real Compass		88
7	Making a Kaleidoscope		101
8	Wooden Toy Train		113
9	A Real Barometer		128
10	Rubber Band Gun		134
11	Whimmydiddle Stick		153
12	Final Comments		158

Chapter 1

Starting Out

Toymaking has always been a part of the fun growing up, and learning to build your own toys from items you find and items that would be thrown away is especially fun. There are all sorts of things thrown away that can be used in making toys-- even paper grocery sacks! So many throw away things can be used in making a toy; the only limit is the imagination of the person who likes to build things. So keep an eye out for anything that you can use, bottle caps, plastic bubble packs like glues and various items come packaged inside, wood scraps, spools, dowels, etc. ... there are so many useful items. Hobby stores also carry quality wood items in various dimensions for the woodworking projects described.

Some of the toys in this book may need a parent's help, others do not. Small woodworking projects can be done with almost nothing, some with a Coping Saw and pocketknife, perhaps a drill and files and sandpaper, but large items might require power saws, use of which can be dangerous and ***should have help from a parent. Always use safety equipment!***

The Coping Saw

A Coping Saw is a very safe saw for a kid to use after a bit of instruction, and it has several properties:

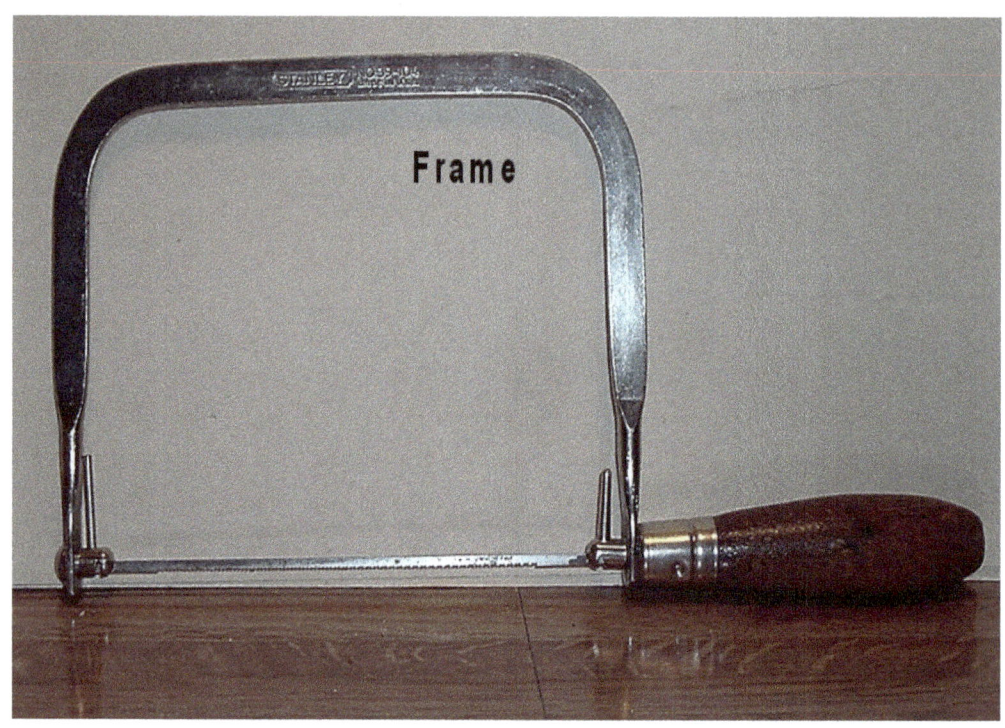

Fig. 1. Coping Saw

- **It is capable of cutting curves.**

- **It can be taken apart, and re-installed through a drilled hole to cut out an inside area.**

- **It has fine blades available for cutting, so does a nicely finished cut on the wood or material.**

- **It has one disadvantage, the blades are thin, and therefore easily broken.**

See Figure 1. Note the fittings at each end of the frame--these allow for taking the saw apart, or aligning the blades. Both ends should be aligned to keep the blade true along its length. The saw blade is easily rotated to allow doing curves. Because the blade is thin, *ALWAYS* wear safety goggles when using the saw!

****

NOTE:
ALWAYS use goggles when using ANY tool, in case a blade or fragment of the work should break and fly off! Protect your eyes! Be safe !

****

Variable Speed Sabre Saw

Another excellent saw for curves is the Variable Speed Sabre Saw, and it is fine for straight cuts as well, but requires some sort of clamping to keep the workpiece stable.

Get some help from an adult to get confident with this saw. Always be sure to clamp so the portion being cut is off the edge of your worktable, so as not to cut into your table. You will likely need to move the workpiece often to keep the cutting area off the edge during cutting. Keep the cord away from the blade.

If you use it for curved cuts, run at a bit lower speed to allow slower cutting and more control; also use a narrow fine blade to allow for nice curves.

****

Always unplug the saw while changing the blade, never leave it plugged in until you are ready to do a cut. Never have your fingers near the blade while cutting! Do not damage your worktable. Make sure the blade does not get near the power cord!

****

Fig. 2. Sabre Saw

Hole Saw

Another extremely useful tool is used with a drill; it is a Hole Saw. It drills an axle hole as it also cuts a perfect wooden wheel! ***When using the Hole Saw always be careful to keep a waste*** *block beneath your workpiece, to protect your work table!*

When using a drill with this saw, always keep the drill unplugged until you have the desired saw blade chucked into the drill. See Figure 3 which shows a single blade mounted in the right hand Hole Saw. This is how you would use it, with the single desired blade. Be careful to drill straight into the surface, and avoid any tilt. ***Don't get your hand or body near the blade!***

Fig. 3. Hole Saw

Belt Sander

Finally, one very useful tool is a Belt Sander, ideal for truing up pieces and smoothing cut marks. With parental guidance, this is an easy and fairly safe tool for a kid to use. It is ideal laying upside down with the belt sanding surface on top, and this allows easy control of small pieces for smoothing. Use Safety glasses and a dust mask for this tool. *Careful with your fingers!!*

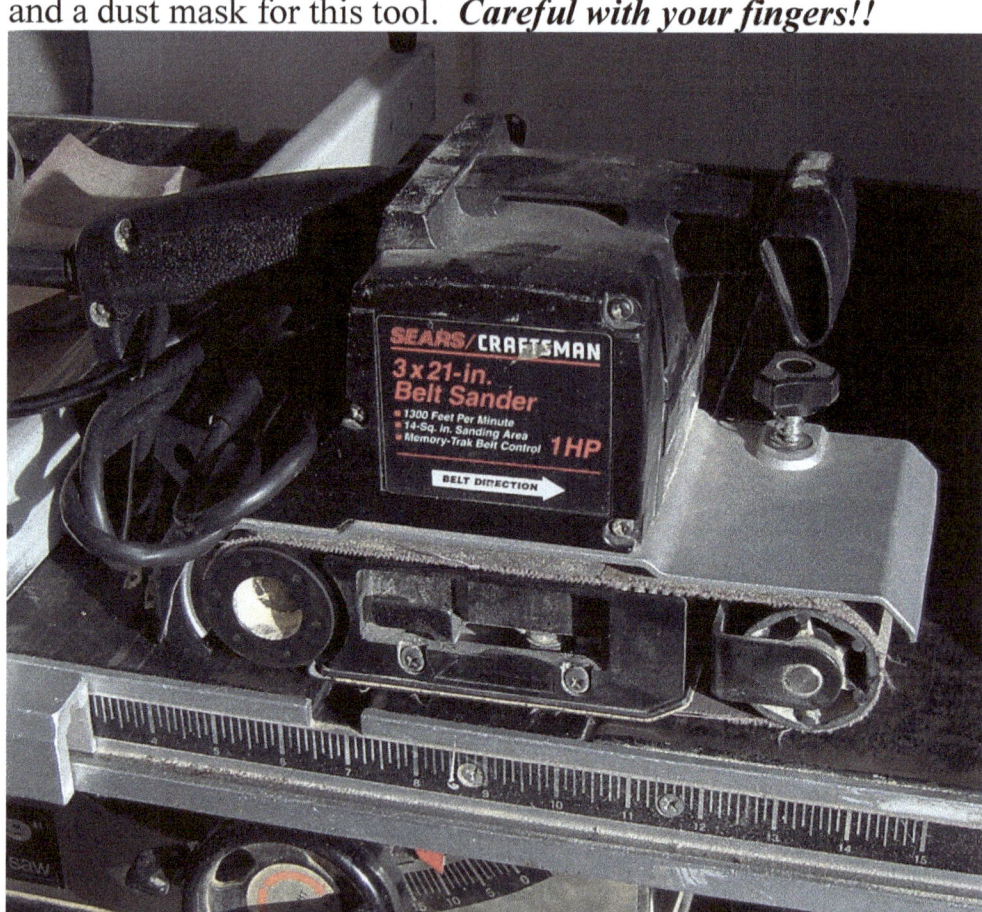

Fig. 4. Belt Sander

When using the Belt Sander, keep a very secure grip on small workpieces and apply light pressure onto the sanding belt as you smooth, and move the piece back and forth somewhat across the belt to gain use of the whole sanding surface, you do not want to wear one particular area. *Avoid your hand hitting the belt!*

If desired, you have many choices of grit too, from coarse to very fine. Depending on your saw cuts, you may need a couple different grits. *Use eye protection and a dust mask.*

*********************************** ***********

Table Saw

A Table Saw is something your PARENTS might use to help you make some parts. **Do not try to use this by yourself as it is EXTREMELY DANGEROUS; this saw should only be used by a skilled parent. It can cut your arm or finger off in an instant with one single mistake !!**

****

Pocketknife and other hand tools

Another tool that is used requires a parent for guidance at first, a pocketknife. Certain simple rules apply with a pocket knife:

- **Never cut towards yourself.**

- **Use your hands braced against each other to limit the movement of the blade. In this way you prevent long swipes with the blade, and allow only short, careful and distinct controlled cuts.**

- **Keep the blade sharp; a dull blade slips more often and is actually less safe.**

Every so often as a kid, I cut myself, my right forefinger has many scars from those accidents. They were very scary at first but after a few times, I became less fearful. You go in the house, wash the finger off and hold it until it quits bleeding, then put on a bandage if you want. And then you go back to work, remembering how you did the nick and being more careful next time! See Figure 5. These are some other small hand tools that are handy for the projects. ***Be especially careful when using the chisel for clearing out excess wood, it is extremely sharp! The Awl is used for marking an indent for drilling. The wire clipper is nice for nipping off stir sticks.***

Glues

There are many glues you can use for projects, some are particularly aimed at woods and paper. These are carpenters' glues. There are also Acrylic types of handyman glues, "Loctite" is one brand, using plastic binders, and these are useful for gluing plastics to wood, wood to wood, etc. "Elmer's," "Loctite" and "Titebond" and several other wood glue brands are found in most stores. There are so many brands of these types of glue, probably all are fine for the projects described in this book.

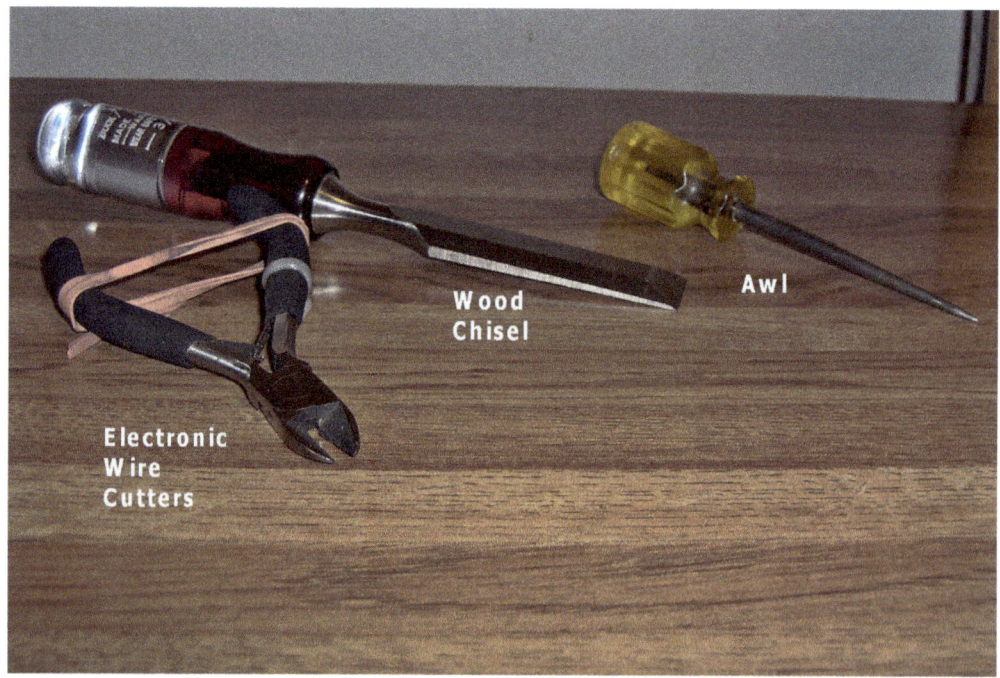

Fig. 5. Helpful Hand Tools

Finishes

There are many possible finishes available for protecting your projects, in particular wood items.

Always sand the surfaces smooth prior to applying the finish.

There are two common varnishes, water bases Polycrylic Varnish and oil based Polyurethane Varnish. These are available in both spray and brush types, both are fine for protecting and finishing toys. These essentially are surface coats.

Another finish is Danish Oil, basically a treated Linseed Oil product which is put on and then after a bit, wiped off to leave a smooth finish. This oil soaks into the surface a bit, rather than just coating it. Natural and Golden Oak are ideal.

Whenever your toys are made of wood, it is a good idea to use one of these to protect your toy. Follow directions on the cans and if spraying, be sure to go well away from any house or auto so the over spray does not ruin a finish. Also you may place your toy down inside a large box during spraying just to add to the safety factor. *Avoid breathing the vapors.*

Safety

In any project attempt, the very first and foremost item to be kept in mind is:

WORK SAFELY !! Before ever taking a tool in hand, think to yourself, "Do I need my safety glasses, or a dust mask?" (Or ear protection- if using power tools..)

Your eyes, your lungs, and your hearing are irreplaceable... so protect them!!

Keep your body, hands, power cords, etc. clear of any cutting, sanding or drilling work. Also insure that your worktable is protected from accidental drilling or cutting. BE SAFE!!!!

Chapter 2

Building a
Little Crawler Toy

Items Needed: Wooden Spool,
Two sticks about match sized,
Plastic Button or small piece of soap,
Rubber band,
Teflon Tape

    This little toy was one made often by kids in the 1950's, and was very fun to play with. Wood spools from sewing were plentiful and common and a small stick , a rubber band and a piece of soap or a button were the remaining items need to make it.

    Basically it uses the rubber band to power it; one stick drags allowing the entire unit to be wound up and placed on the floor or carpet, the dragging stick lets it power along like a little tractor. With the fun of building a little domino fort or some castle wall, the little device acts like a army tank, ramming and knocking down the dominoes . Or it is also nice as a little racer for competitions.

    In the old days, a small piece of hard soap could serve as the "washer" but a smooth slick button or bottle cap with Teflon

Tape also works. The hole must be made large enough to allow the rubber band to go through and loop over the long matchstick.

It is easy to make a hole in a bit of soap, but it is difficult to find the hard soaps used back in the 50's, which worked well, so consider a button, it may take a careful job with a drill. So this must be done with care! *Adult Supervision is advised unless you are used to using a drill.* You could hold the button in a clothespin flat for drilling. Or you can trim a bottle cap into a small washer with the Coping Saw. Be safe. **Use safety glasses!**

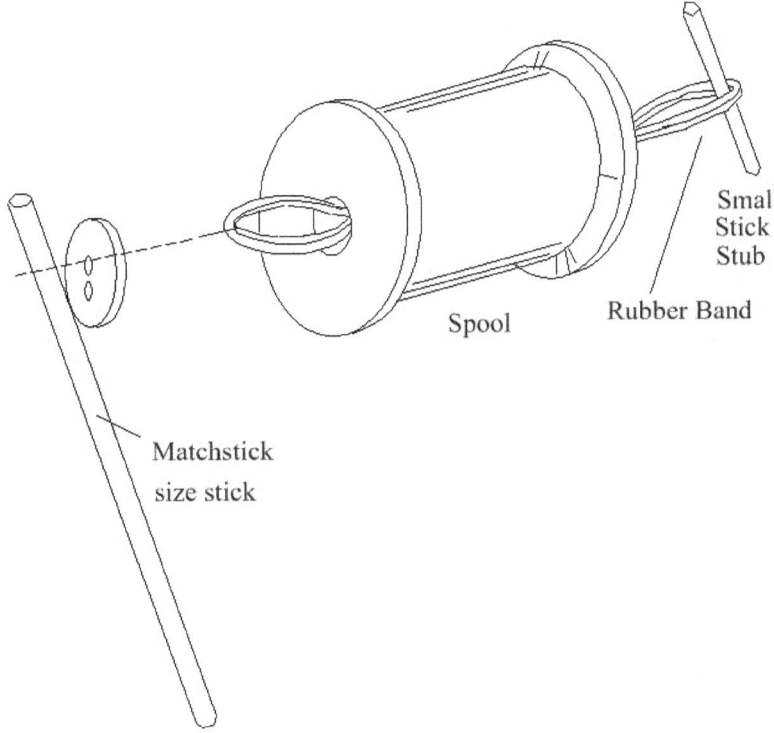

Fig. 6. Spool Crawler Toy

The rubber band should be a bit longer than the spool, and the small stick stub can be attached to the spool with a bit of wax or tiny spot of glue to prevent slipping but still allow removal if the rubber band needs replacing.

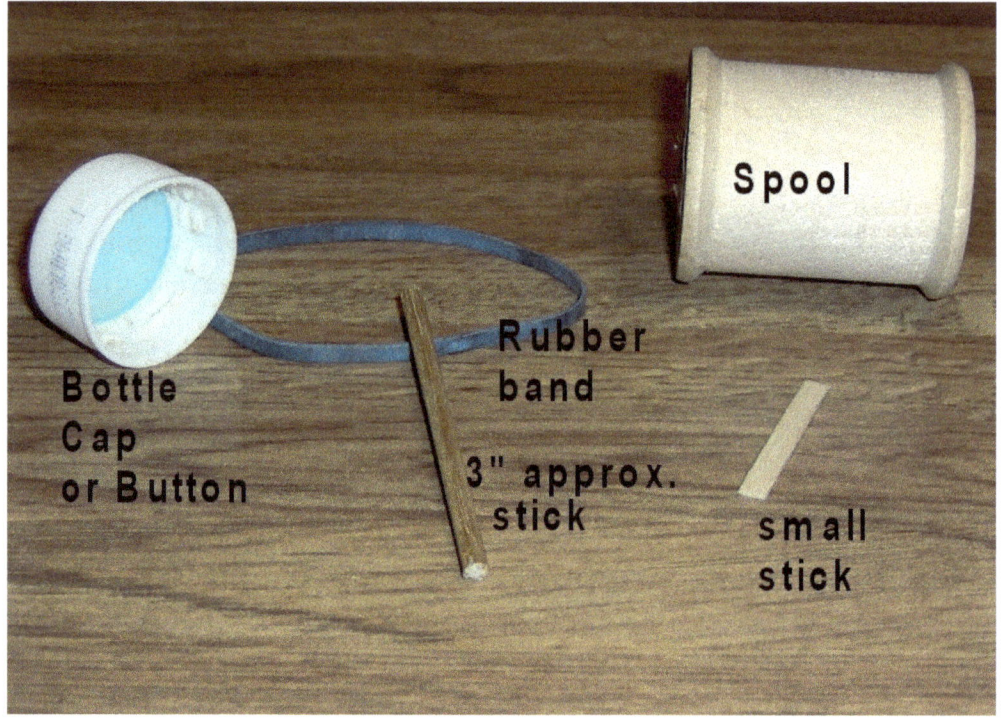

Fig. 7. Parts to Make Spool Toy

Cut the collar off the bottle cap and drill a hole in the center to accommodate the rubber band, probably 3/16" ….or drill a hole in the button to form a washer. If using the bottle cap, reduce the size too. Trim any snags. Glue on a bit of Teflon Tape with Loctite Handyman's glue.

Assemble the toy to look as in Figure 8. The Teflon should rub against the spool side. Using the long stick, hold the spool and wind the stick several turns. Holding the spool and stick locked under tension set down on the floor. The toy will zip across the floor, a fun, great and simple little gadget, lots of fun for assaulting a domino fortress!

Fig. 8. Finished Toy

Chapter 3

Two Army Tanks

<u>Items Needed:</u>   Scraps of wood, 1/4", 1/2" & 3/4" thick,
1/4" and 1/8" Dowel,
#4 Box Nail

The little Army Tank described here is like one I made in the early 1950's when I was perhaps 12 years old, living in The Dalles, Oregon. It is made from a small block of wood and a few scraps. It was made using a Coping Saw and pocketknife, possibly a handsaw to cut it from a scrap board.

The little tank is similar to a real one, the M-3 Stuart Light Tank from the days before and during World War 2. It was a light tank, small, and had a 37 mm cannon and machine guns. The British used it in North Africa.

See Figure 9. Look this tank up in the library or on the internet and read about it.

****

When you see something interesting in books or on the internet, try thinking of a way to build a model of it or learn its principles or history or in the case of other subjects, Electronics, Mathematics, Algebra, Science, Submarines, Airplanes, Physics, Magic, etc. ...whatever; try to learn as much as you can! Even without school you could learn so much from a library or the internet to become knowledgeable in virtually anything which interests you! Learning is endless and fun, there are so many things to learn about! You can become an expert in whatever you wish.

****

Fig 9. Stuart M-3 Light Tank (From Archives.)

The Stuart Tank was not adequate in North Africa during the war except as a scouting vehicle; it was too light in armor and the cannon was too small for fighting against heavier German tanks. However it was useful in the Pacific Islands as a jungle tank, it was reliable and rugged for that purpose. And it was actually used in some foreign countries right up into the sixties. A pretty amazing life for an Army tank designed in the 30's. Read about it on the Internet.

Now look at the little model I made in the 1950's on the next page. It is very small. But you can see it bears a close resemblance to the real M-3 Stuart Tank.

I saw a picture in some magazine or article and made the model from studying that picture. The same method you can use to make a model of your favorite thing.

The little marks on the turret were from the old model in my picture that had firing ports in the turret.

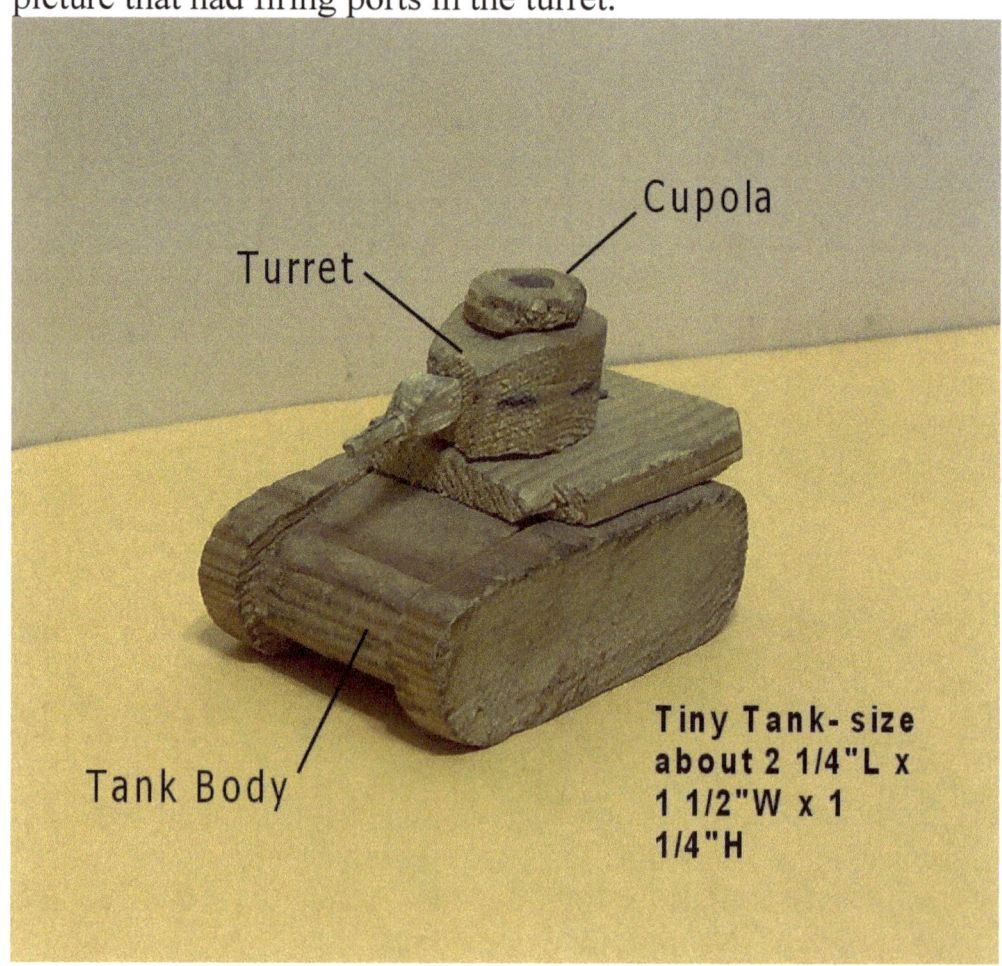

Fig. 10. My Actual Tank, Made in the 1950's

Looking at the little tank, you will see it is made of five pieces, the body is a single block; there is a thin piece above the tank body, then a turret, and a small piece above that, a cupola. The cannon is the last piece. The main body part was a difficult piece to make the way I did it... cutting across between portions that were the tracks. It is done by cutting along each side about 5/16" in from the sides, then using the Coping Saw to saw across and form the sloped front. It required using the coping saw and then additional cutting with the pocketknife to finish the slope fairly evenly and flat across between the tread portion. The turret, the cupola and the thin piece were done mostly with the Coping Saw.

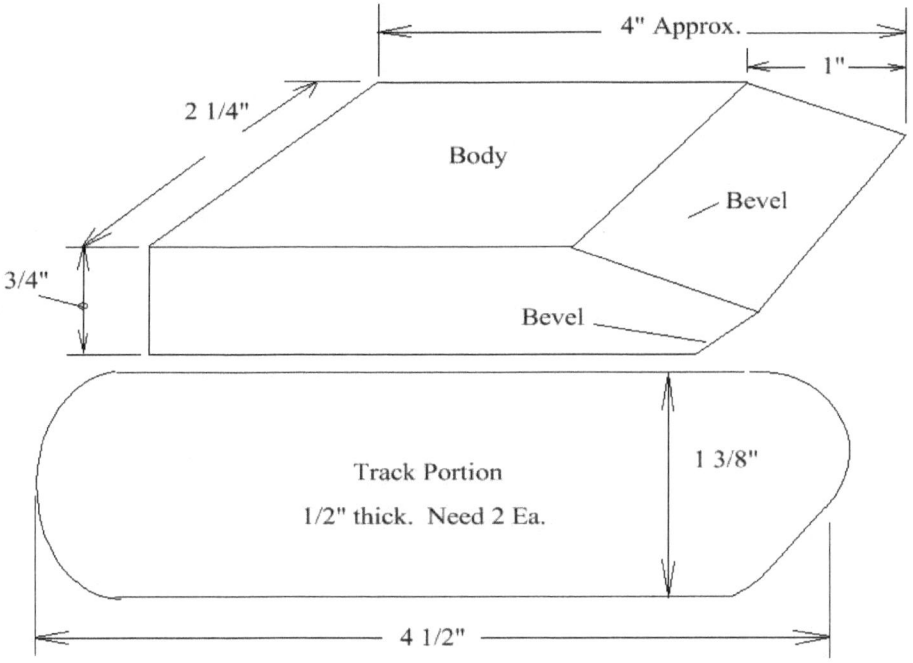

Fig. 11. Body and Track Portions

It doesn't take very much material to make the little tank: a small block of 3/4" pine or similar. And some pieces of 1/4" thick stock. If you study Figure 10, you will see the tank body grain runs crosswise looking from the front on the main block; with a soft wood, this makes it fairly easy to cut the angular sloping front of the tank with the Coping Saw, leaving a tread portion on the two sides.

Fig. 12. Cutting the Tracks

The main requirement for doing it this way is soft wood like pine, and straight grain.

Luckily, for other woods, there is also a much easier way to make the tank, by using separate pieces for the tracks, see Figure 11. This is a slightly bigger tank, and therefore easier to build. This is the first one we will do; it will be fine as a project. You can cut out the tracks best with a Sabre Saw or Jigsaw, but if it is not available, a Coping Saw can cut the curved ends. The tracks are cut from a 1/2" thick soft pine piece.

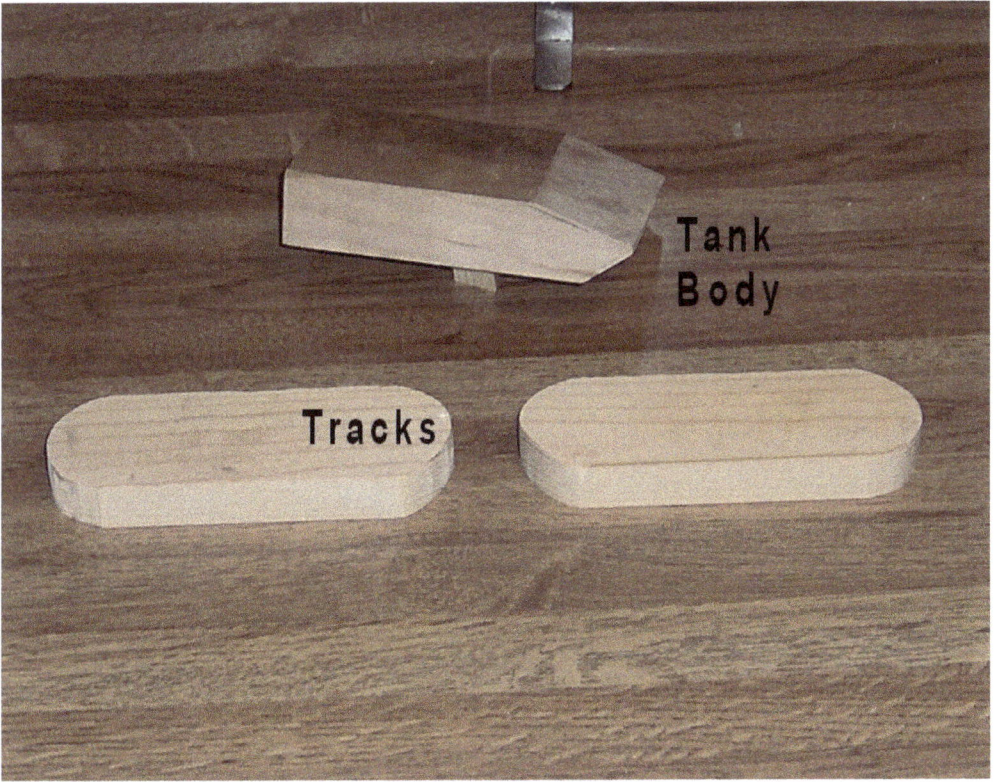

Fig. 13. Finished Body and Tracks

If you can cut them both together, they will match fairly close and once sanded you can true them nicely to match. See Figure 12. Then cut the beveled body portion. Sand and file as necessary to smooth it.

Once sanded, the parts will appear as in Figure 13. After gluing the parts together you will have the result shown in Figure 14. (**Note the safety goggles … always wear them when working.)**

Next comes the thin piece above the body. With this scaling, 1/2" thick is about right. Cut a piece to fit on the upper level portion of the tank, width is slightly less than the body, about 2 1/4" wide, length about 2 3/4".

Fig. 14. Body With Tracks Attached

Cut two 3/8" pieces out of each front corner; this simulates the real tank body machine gun setup, as shown in Figure 9.

You can glue on this piece next to complete the body portion. See Figure 15. Note the clamps used for holding the parts while gluing. These are a common type of clamp usable for small clamping jobs; larger jobs will perhaps require bar type clamps. These type of clamps are available from hardware stores and suppliers, Home Depot, Lowe's, and Harbor Freight for example.

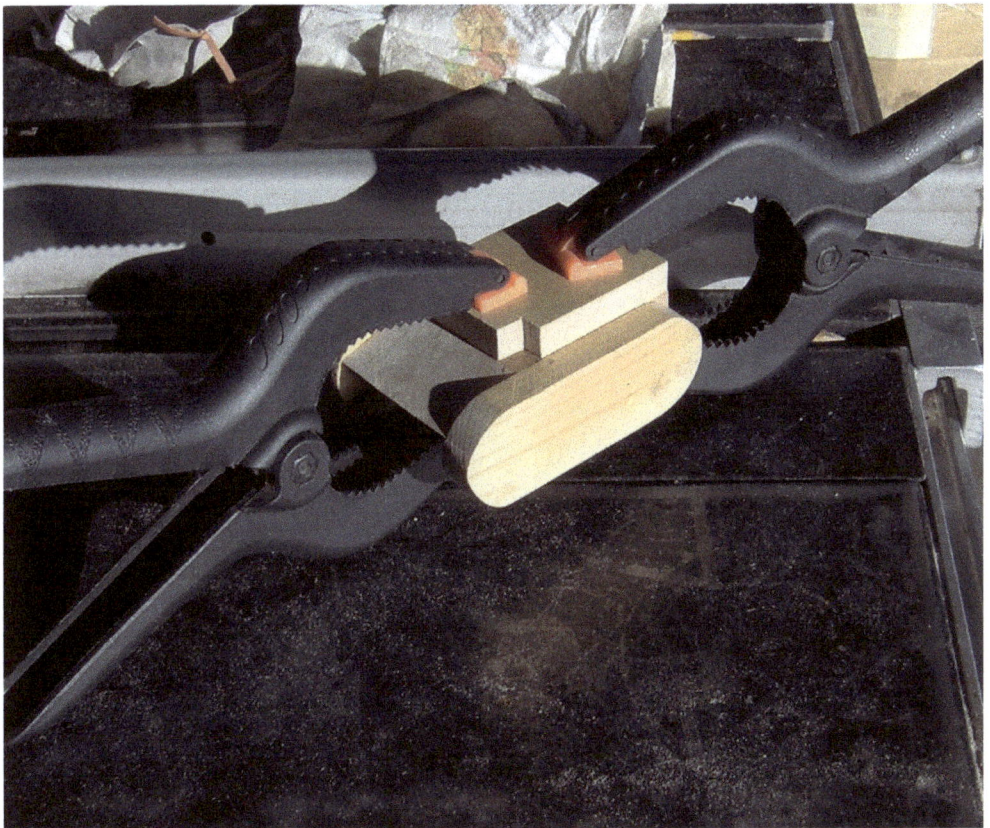

Fig. 15. Clamps Gluing Parts

Look at Figure 16, which shows the turret piece. Cut this from a 3/4" or slightly thicker piece of wood, and also cut a small cupola piece from 1/4" material, either round or Hexagonal (6 sides) as shown in Figure 16.

Cut or sand a slight bevel onto the front of the Turret. Glue the Cupola onto it, position as you wish, trying to come close to the picture in Figure 10.

Fashion a barrel as in Figure 16. The larger portion can be left off and drilled to fit over the 1/8" barrel later to finish the look.

Drill a 1/8" hole centered into the Turret for the barrel, and glue it in.

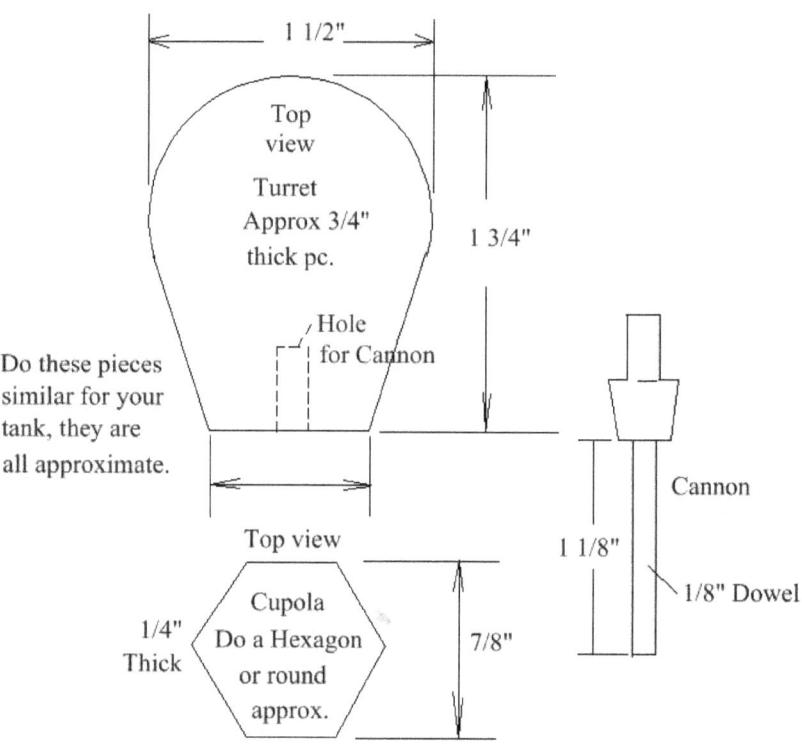

Fig. 16. Parts for Turret and Cupola

Now, if you wish, you can drill a centered hole through the tank body and a bit into the Turret. If you use a 1/4" drill, you can then glue a 1/4" dowel into the Turret long enough to come down through the tank body . Enlarge the hole in the tank body to 5/16". then the dowel can come through the tank body and be glued into a small block of 1/4" material drilled to 1/4" to allow the turret to turn freely.

See the described parts in Figure 17. The final tank is shown in Figure 18 . The Turret is moveable, and it is a fairly good facsimile of the real Stuart M-3 Tank.

Fig. 17. Making the Turret Moveable

You can add the recoil collar over the barrel at the turret if you wish, and other machine guns that are on the real tank to make it more real, also the model can be painted.

If you have gotten this far, you have completed a large version of the tank. Danish Oil will provide a nice finish.

Fig. 18. Finished Larger Tank

Building a Small Tank

If you wish to do a small tank as I did in the 50's, you need to start with a 3/4" thick soft pine board with width of 2 1/4" and it must have a very straight grain, no knots or twisty grain.

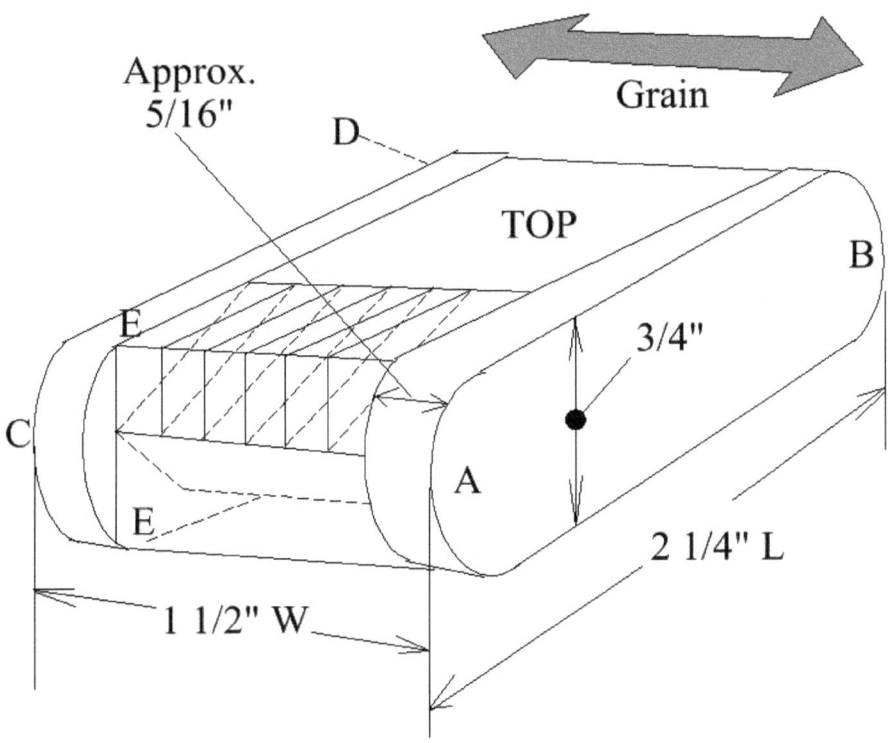

Fig. 19. Showing Cuts

Do most of your cutting of the front bevels, and other portions **before** cutting the little tank body piece from the board. This allows you to hold the board more securely while doing most of the cutting.

Mark the points where the track portion will lie and clamp the board in a vise or on a workbench to hold it steady. Then do multiple angle cuts to form for the front top sloped portion of the tank, see Figures 19 through 25.

Then cut along each side as shown in Figure 23 and 24. Also do the angled cuts for the front bottom slope of the body.

Fig. 20. Initial Cuts

By careful splitting off of the front angle waste pieces and smoothing with the pocketknife and sanding, you can achieve a smooth sloped front.

The underneath portion is cut on each side with the inner waste removed to form the treads, and underneath belly portion of the tank.

Fig. 21. Removing Bevel Waste

It is a good idea to do several cuts close together so you can carve or *chisel* out the portions in between more easily.

A Chisel is ideal for cleaning out the inner material, however, **the Chisel is a very sharp, dangerous tool, doing this should have the help of an adult. Do not attempt it without help! A serious cut could occur, so ask for help.**

Fig. 22. Bevel Shaping

Turn the piece over and do the bottom front slope. These multiple cuts will allow easily splitting off the small sections and shaping the sloped portion.

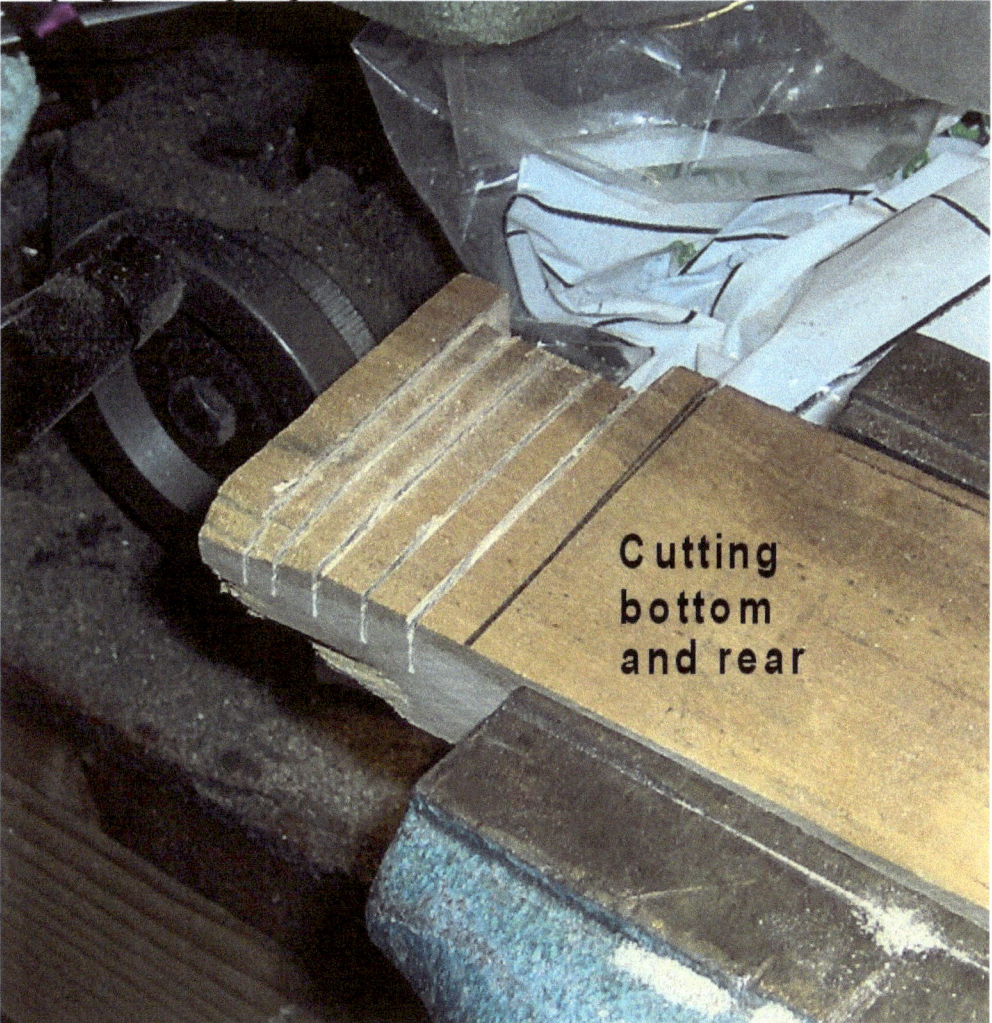

Fig. 23. Bottom Cuts

They should go back a bit into the front to allow for the track portions on both sides to protrude a bit in front of the tank body. To do the other portion of the flat bottom, continue with the piece turned upside down, and this time cuts should go into the wood clear from the front to the rear of the little tank body.

Fig. 24. Removing Waste

Cut into the rear of the tank with slight bevels, perhaps 1/4" or so. These are needed to allow some shaping of the treads at the rear of the tank.

Fig. 25. Using Chisel

At this point use the wood chisel (or pocketknife to clean out the waste wood and leave the tank body ready for sanding.

After the bevels and bottom look fine, cut the little tank body from the board. You will have the body pretty ready except for some minor splitting to form the round track portion front and rear. This is not hard as we picked a soft wood with the grain crossways.

The turret and other parts are done with a Coping Saw in about half scale of Figure 16. The piece above the body should be about 1/4" thick, and can be cut from a small piece of 1/4" plywood. The piece should end up about 1 1/4" x 1 1/2" . Nip the two small corner pieces out with the Coping Saw. Set in place and glue to the body just behind the front bevel.

Next, draw the outline of the turret out on a piece of 3/4" wood, and before cutting it out, drill a small barrel hole in the Turret. Then clamp the large piece in a vise and cut out the Turret. Sand it and cut or sand a small slight bevel on the front, top portion. Use the Coping Saw to remove about 1/8" from the bottom, as 3/4" is a bit too thick for this scale. Figure 26 through 28 show the turret.

Cut the cupola from some 1/8" material or a similar scrap. The parts are quite small, and it is fine to just glue the cupola to the turret. A 1/8" dowel or nail can be used to mount the turret assembly to the tank body. I used a small box nail in my 50's tank and did here too, but drilled through the turret unit first with a 3/32" bit and through the body with a 5/64". After all this work it would be terrible to split the tank.

Also chip the corners off the track areas and use sandpaper or the Belt Sander to carefully finish rounding the tracks front and rear.

Once you are done with this, gently tap a #4 Box Nail through the turret assembly into the body. You will be able to turn the turret, but it will be locked nicely to the tank body.

Use a small scrap of solid wood about 1/8" thick and cut a small barrel with a pocket knife and the Coping Saw. See Figure 28 for an idea of the desired result. Round the barrel portion.

Glue into the turret.

Fig. 26. Cutting Small Tank Turret

Fig. 27. Small Tank Turret With Cupola

Note the front bevel on the turret; you can see it better in Figure 28. Use the Coping Saw to thin it slightly top to bottom as 3/4" is a tad too thick. Note the cupola is simply glued on to the top. Sand it to finish.

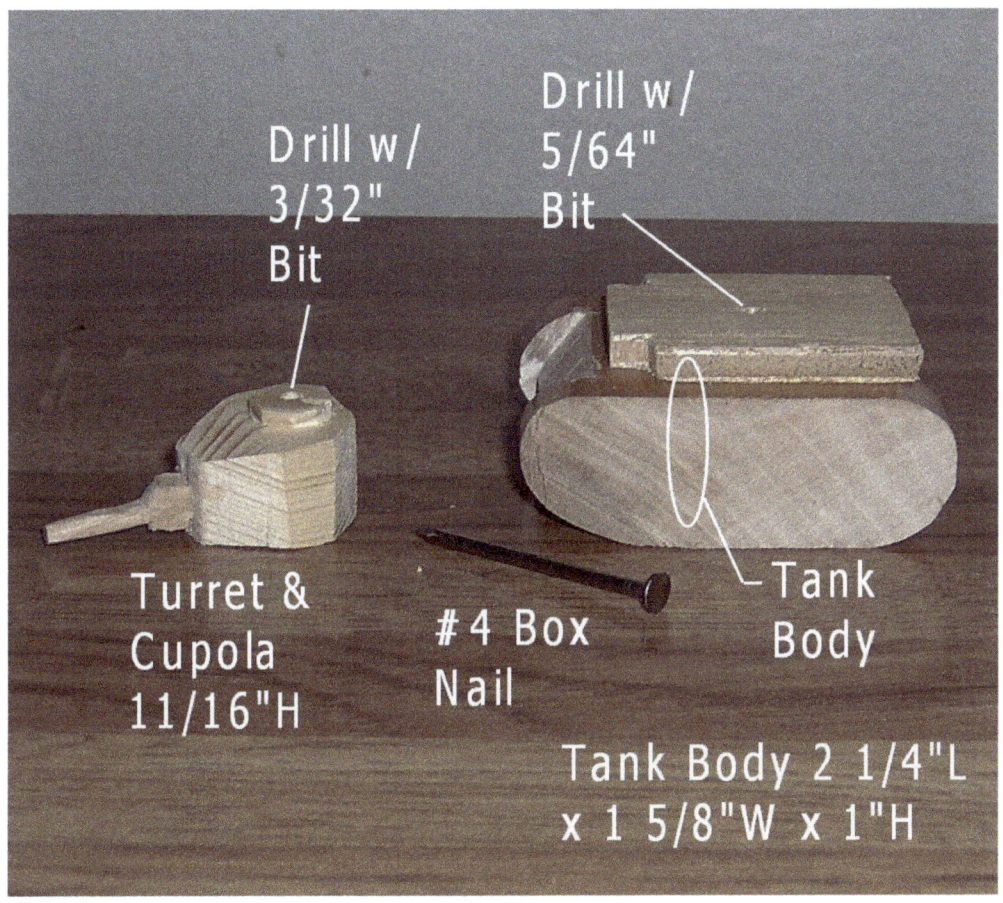

Fig. 28. Tank Parts

You can finish the little tank with Danish Oil if you wish .. to provide sealing and protection for the wood.

The finished small tank is shown on the next page.

Fig. 29. Little Tank Finished

Chapter 3

Historic Airplane Built From Scratch of Coffee Stir Sticks

<u>Items Needed:</u> Coffee Stir Sticks, Carpenters' Glue,
Hobby Clamps or Clothespins,
Rubber Bands, Grocery Sack,
Plastic for Canopy, Baling Wire,
Hobby Paint,
Small block of wood for Propeller Hub

    This chapter shows how to build a very nice airplane of coffee stir sticks and a paper sack, with some bubble pack clear plastic as the canopy. The structure is made of the wooden stick framework and is very much like a real airplane looks beneath its skin.

    In the beginning of the last century, real airplanes actually used wood construction beneath with a fabric covering, this was particularly common in biplanes of that time, during World War 1, and in early mail planes and "barnstorming shows" by daredevil pilots. These pilots flew about the country doing airshows and such, thrilling patrons with daring wing walkers and similar stunts.

    As a kid, I built a model like described here, made from split off pieces of a fruit crate, and covered in paper. It was a biplane model, and I took a picture of it on a concrete patio with a warehouse across the street in the background to make a trick photo as if it was on a runway with hanger behind. (This was in Brigham City, Utah in the mid 1950's...) At the Brigham City Library I had been reading a book about a journalist using a trick photograph in reporting the sinking of the Battleship Maine in Havana Harbor, a trigger for the Spanish-American War. That was the thought behind doing my picture with the warehouse to simulate a hanger.

My Mother noted on the back of the photograph she had a doily the same color... I had spilled the paint on it while finishing the outside. Here you see a view of my old photograph in Figure 30, next page.

The framework was covered in thin paper, and painted with hobby paint. You can see that with paint, the model can be very neat. The wheels were cardboard with painted tires.

It was tricky and time consuming to split thin strips of wood from the fruit crate as well as cutting bulkheads for supporting the fuselage portion. Nowadays there is a much easier way, using coffee stir sticks of wood. These come in a box of 1000 for only three to five dollars from office supply stores, like Staples or Office Depot.

A hobby store, or other craft stores carry small plastic hobby clamps which are ideal for clamping parts while gluing, and you could also use clothespins as a substitute. I would recommend plastic types if you use clothespins as they won't stick to the wood glue.

This type of airplane takes dedication to build, it isn't something you do instantly, but the reward of the finished version is well worth it. You can build a fabulous airplane!

Fig. 30. Model Plane made in 1950's

Fig. 31. Archive Photo from Google,
Bell P-39 AiraCobra

The model plane described in this chapter will be a P-39 Airacobra from WW 2. The aircraft was very innovative; it was the first aircraft with a tricycle landing gear, and the engine was **behind the cockpit,** totally different than other aircraft. It also housed a 37 mm cannon firing through the propeller hub. It had two .50 caliber machine guns in the nose and two .30 caliber machine guns in each wing. Very tremendous on firepower!

It was originally built as a top line fighter and did have excellent characteristics initially but during testing a decision was made to cut out one stage of the supercharger. This unfortunately made it unusable as a high altitude fighter, however it served well as a low altitude fighter and ground support aircraft. It achieved more kills by the Russians than any other WW 2 aircraft according to Wikipedia sources.

The next photos will show the step-by-step building of an approximate model of the Airacobra. The builder can pick out any aircraft to do a model, the general construction is identical other than trying to "copy" the look of the desired plane.

In looking at various pictures on the internet or in a book of aircraft, there will be some pictures which give a good look at the wing shape, and this is a good place to start. Also the builder can decide how large to make his model, to keep it proportional, etc. as the parts are fabricated.

Some Pointers on Construction

- Using coffee stir sticks, lengths can be easily broken to size using a finger nail as the break off guide.

- For curved pieces, soaking a length of stir stick in water makes it possible to easily bend the pieces -even into a full circle! These make very strong bulkheads when the ends are glued together to form a full hoop.

- When a stick is wetted for doing a hoop, move around the stick shaping the bend in 1/8 inch increments using a fingernail as the bend point, and bend each increment about 45 degrees as you move around. After a bit of practice you will get the hang of it.

- Use clothespins or small hobby clamps (even electronic clip leads...) to clamp parts while gluing.

- Use your pictures gleaned from the book or internet to guide you in shaping parts to the approximate ratio to make your model look close to the real airplane. You can get a good approximation by eye to do this.

To begin a wing assembly as well as members for the fuselage, first glue three stir sticks together end to end (if using the 5 1/4" sticks) or two together if using 7" size stir sticks. Allow about a half inch overlap, apply glue and clamp. Make at least eight of these for the initial fuselage of the airplane, and maybe ten more for starting the wings. Select nice straight sticks for these, and be sure to keep them straight.

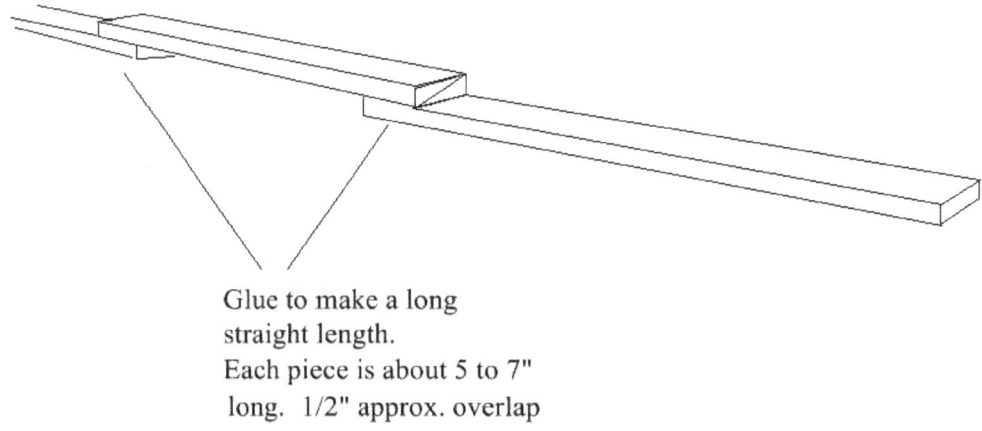

Glue to make a long straight length. Each piece is about 5 to 7" long. 1/2" approx. overlap

Fig. 32. Making Long Spars

Generally, a real airplane consists of a fuselage and wings, and a tail rudder and stabilizers. There are various spars and stringers which join the whole system together, along with bulkheads here and there to add support and strength. All this rib assembly forms a framework which is covered with a skin to complete the airplane. See Figure 33 .

This is how a real airplane structure basically looks inside.

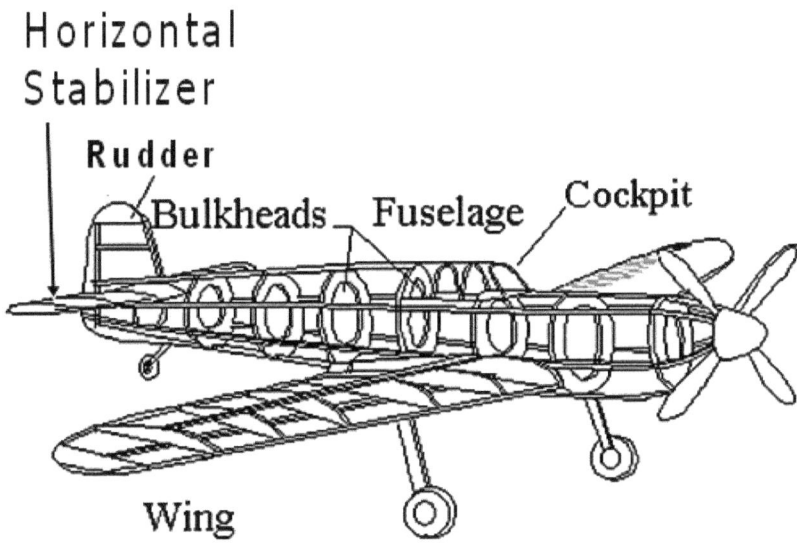

Fig. 33. Basic Airplane Structure

Our model will follow this sort of general look, using the stir sticks, some forming the long strips along the fuselage and wings, some wet and bent to form the bulkheads and curved wing portions.

Look at Figure 34, which shows a mostly finished wing assembly. To start the wings, keep in mind some general principles for a reasonably sized model airplane: the wings should be about 9 to 10 inches long, the width at the widest portion should be about, say about 4 inches. A typical wing tapers to the end somewhat so look at the real airplane picture of your chosen type, and estimate how it should look, and measure to get an approximate width at the end, so the proportion appears realistic. Also note the approximate curved portion created by gluing pieces to form a rounded end. The wingtip of the model shown is about 2 ½ inches across out near the small end, that gives you an idea.

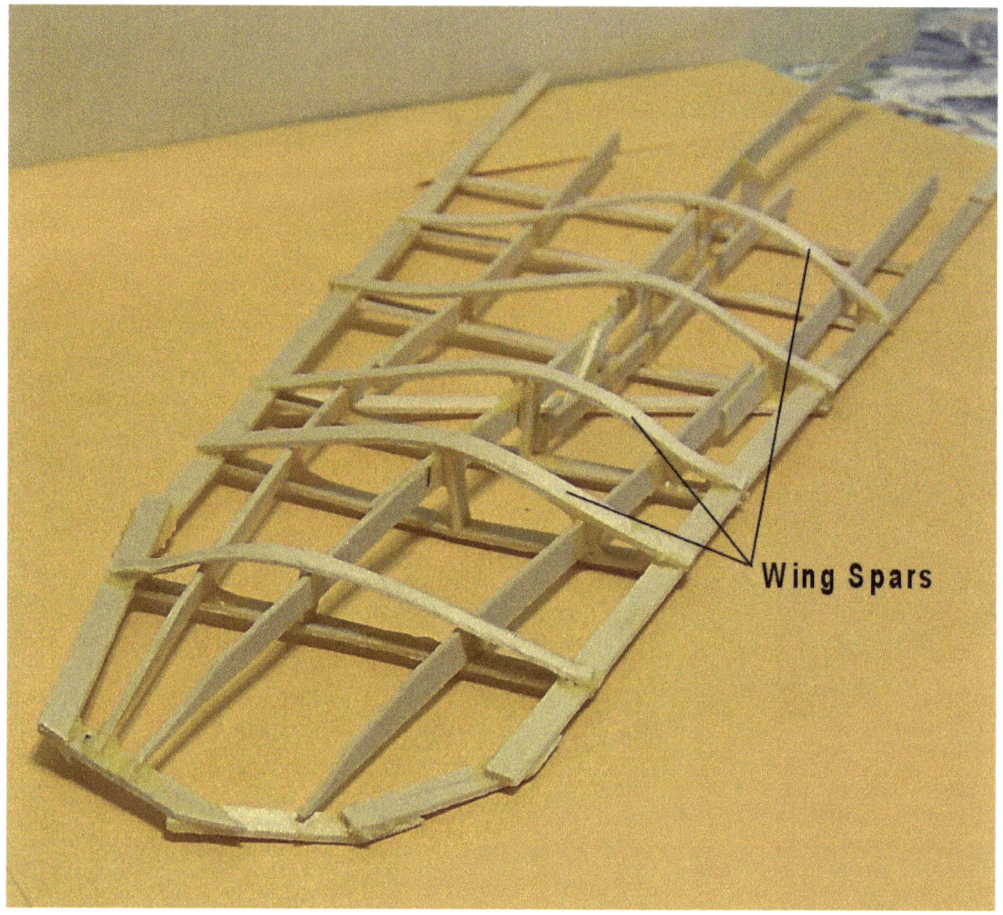

Fig. 34. Basic Wing Shape.

Starting the Wing

To start a wing, lay two long strips out on a newspaper on your work area. Always use a newspaper or some such protection for your table or surface. You do not want to drip glue onto a treasured dining room table!!

Fig. 35. Beginning the Wing

Space them about 4 inches apart at one end and 2 ½ inches apart at the other, the span between these two measurements should be about say 9 to 10". (The length of the wing...)

At equal intervals of about 2" apart along your two long strips, glue and clamp a stirring stick across the two long strips. As you do each one snap it off at a suitable length to form the wing taper from the wide to the narrow end. See Figure 35 for a general idea. Do only one wing portion for now.

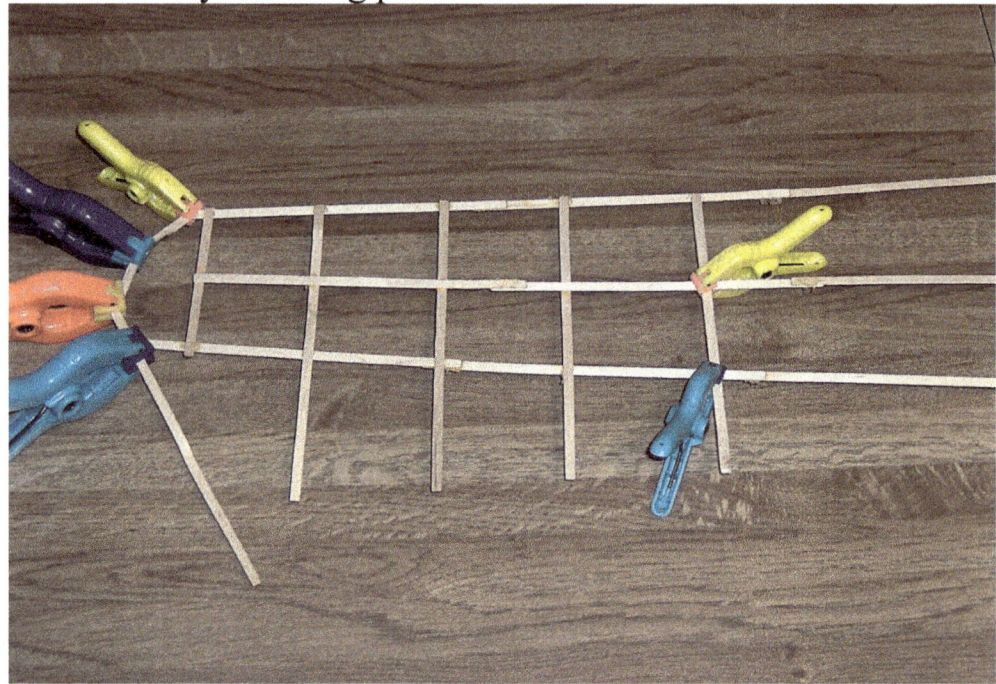

Fig. 36. Middle Long Spar and Tip of Wing

Once you have the entire wing with suitable cross spars, run a center support strip the length of the wing, and then turn it over and do a second wing assembly laying over the back as a mirror image. You want them to match but be for opposite sides.

After you have a second basic wing shape, it is time to add a long spar or two down the length and then begin building the curved spars that form the upper wing shape. Each item you add will strengthen the assembly.

Now is a good time to make a forming jig to shape the upper wing spars to the wing curvature. Then you can wet several stir

Fig. 37. Wing Contour Jig

sticks to have them forming in the jig as you do the rest of the wing. Nail three # 4 Finish Nails into a wood block as shown in Figure 37 to make the jig. If you wet soak several stirrers for 5 minutes in water, then place them in the jig as shown, they will dry to a nice curve for the upper surface wing spars. The jig will hold several spars at once. (Figure 38.)

Look at Figure 39. This shows the main wing girder spar that makes the wing strong over its length. It is sort of like a girder with small triangular braces to make it sturdy. It is mounted perpendicular to the lower wing surface spars, and runs the

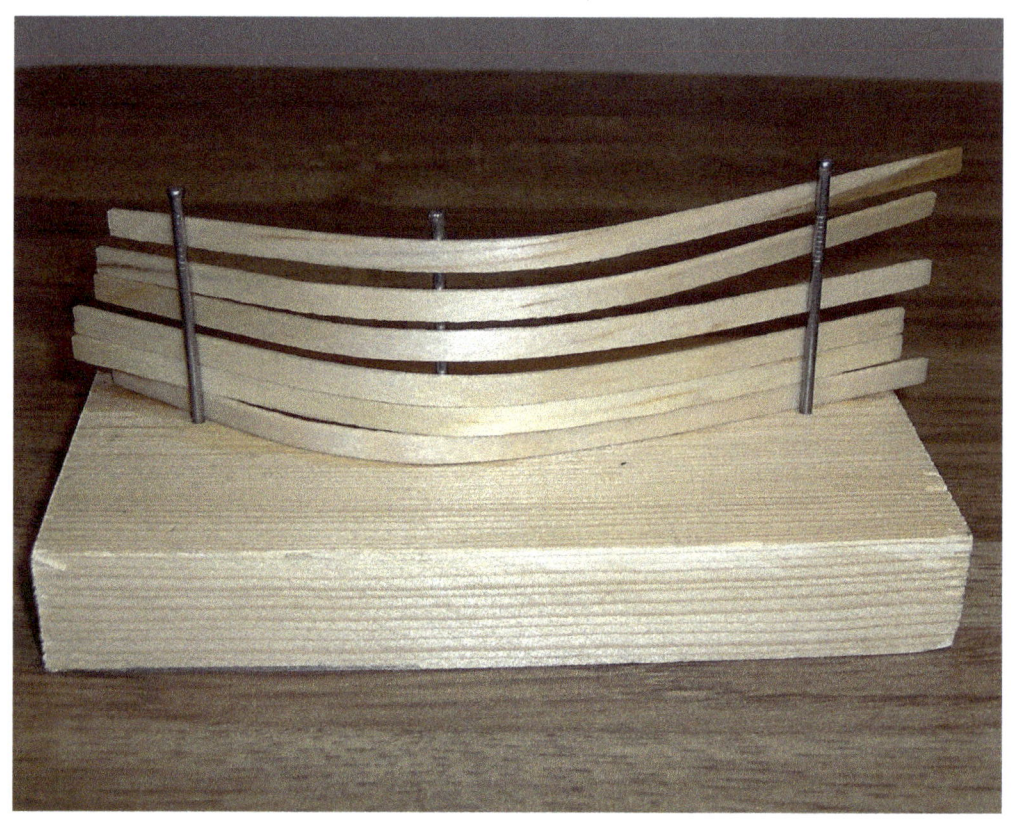

Fig. 38. Spars Being Shaped in Jig

   length of the wing. Note that the girder spar assembly has a top and bottom span, plus cross members and triangular braces, to make it strong. In order to mount this piece vertical to the lower wing surface, use a little trick to apply the pressure to hold it in place as it glues. Use a clamp at the front and rear edges of the wing with a stir stick formed in the jig across the girder to press it down. See Figure 40. Mount the spar more to the front of each wing as it will support the curved spars for the upper surface at the highest point on the wing, which is about 1/3 of the wing width from the front.

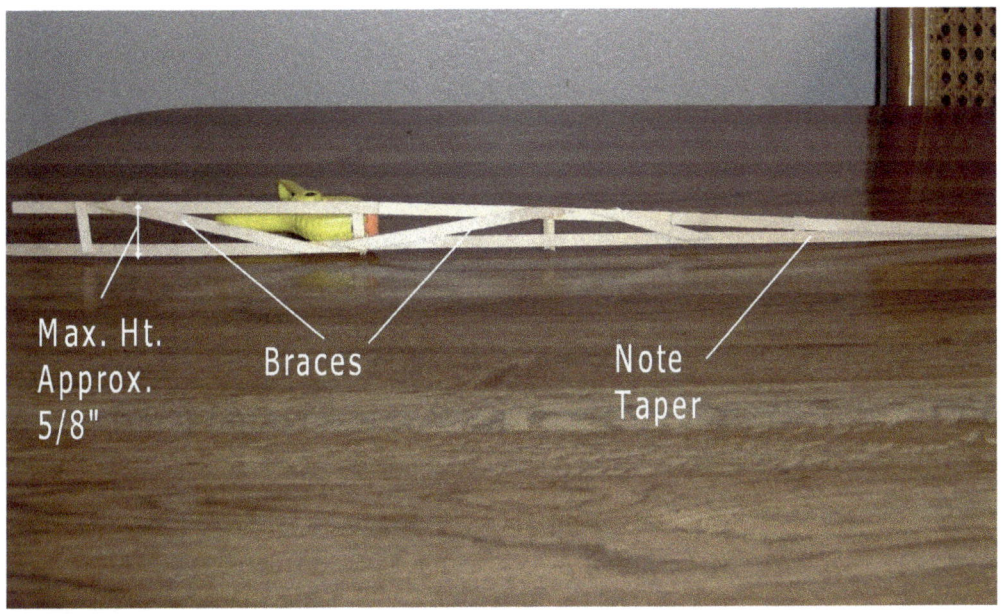

Fig. 39. Wing Girder Spar

    The outer tapered portion of the spar glues to the outer rim of the wing. Once the glued spar has dried, it is time to put some triangular braces in to lock the vertical position of the spar. This stiffens it immensely so that when the curved upper spars are glued across, they will make the wing quite strong, yet still light.

    As they dry, you can start mounting the curved wing spars also. The wing will now start looking as in Figure 41.

    Note: In attaching the front portion of the curved upper spars, the clothespins actually work better than the hobby clamps, which do not handle the steep angle as well.

    Triangles support any member to make it stronger in the plane of the triangle. And any width of a piece acts as a triangle "group" just like a plywood panel supplies support along a wall or in a corner during home construction. Figure 42 explains a bit better than words. Imagine each stir stick as a narrow long version of the picture, so that gluing a "L" shaped portion into

Fig. 40. Curved Spars Helping Gluing

the wing girder spar makes it strong laterally as well as vertically. Of course this is not a huge strength as the width of each stir stick is small.

Later when the paper sheath is added over the wing, there will be a significant strength gain. The paper covering skin acts just like a plywood sheet in home sheathing, adding a huge factor of rigidity. Even the paint will add some strength.

Note Figure 40 shows the left wing; Figure 41 shows the right wing.

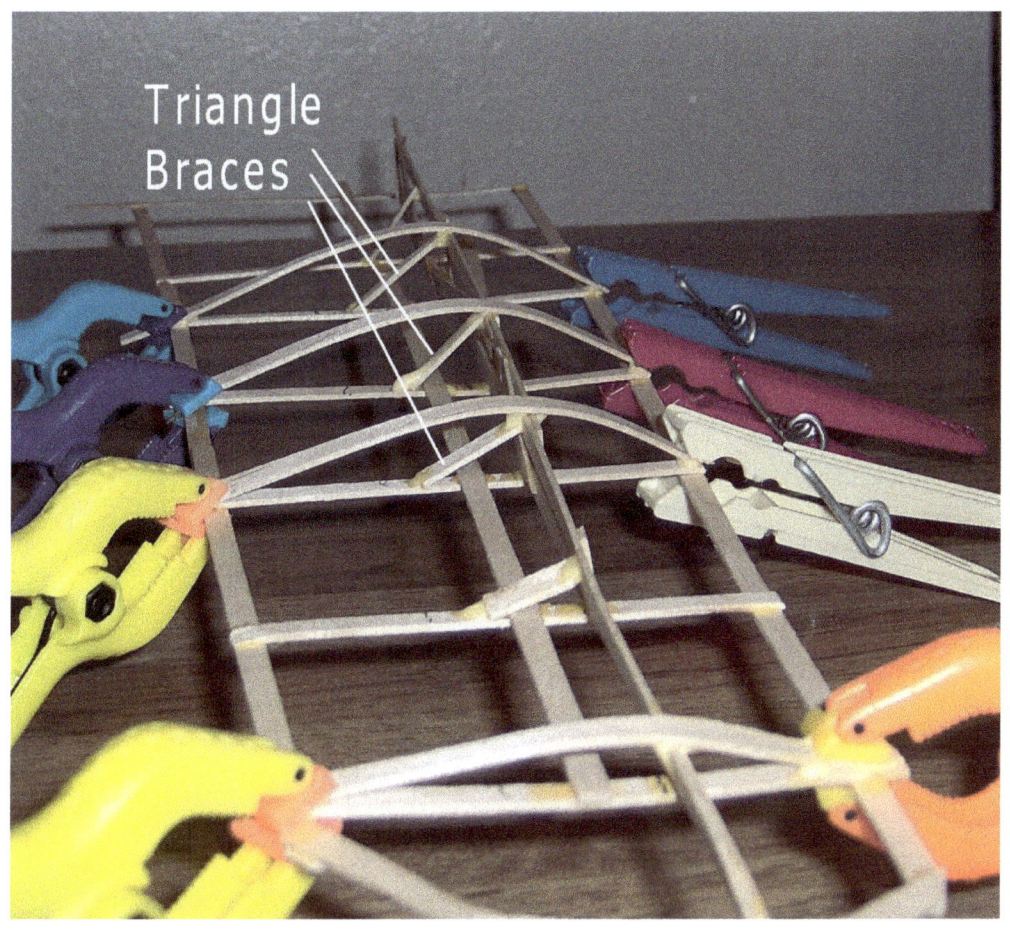

Fig. 41. Triangle Bracing for Main Spar

   Note that clothespins are better for pinching the curved upper contours. The hobby clamps slip off the angle front portion very easily.

   At this point you can also begin gluing some of those upper curved contours on as well. Note the long girder spar running the length of the wing. It is just to the right of the triangle braces. It is glued vertical to the plane of the wing.

   Next we can add a flat spar glued to the top of the girder along its entire length.

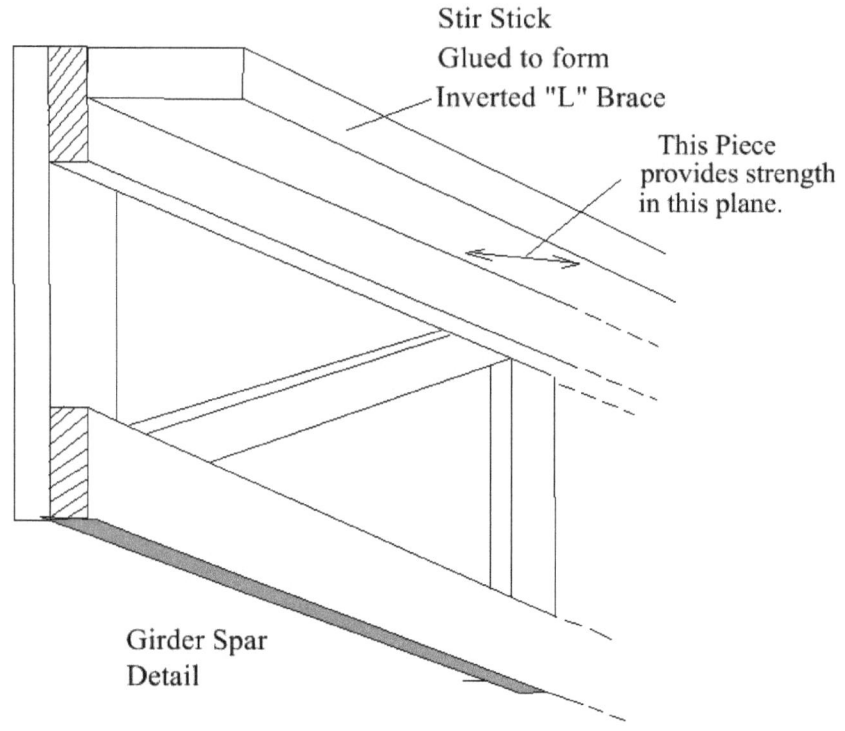

Fig. 42. Magnified Girder with inverted "L"

This spar will duplicate the inverted "L" method to help stiffen the top of the wing.

Also add several "X" braces across the flat bottom surface of the wing, each will stiffen it even more. You will feel an amazing strength as these begin to bolster the wing rigidity. It will still bend but will be quite strong.

Once your "X" braces are glued in, the wing is finished. See the final version shown in Figures 43 through 46. You will see the "X" members on the bottom, the inverted "L" is glued along the top right side of the long girder spar.

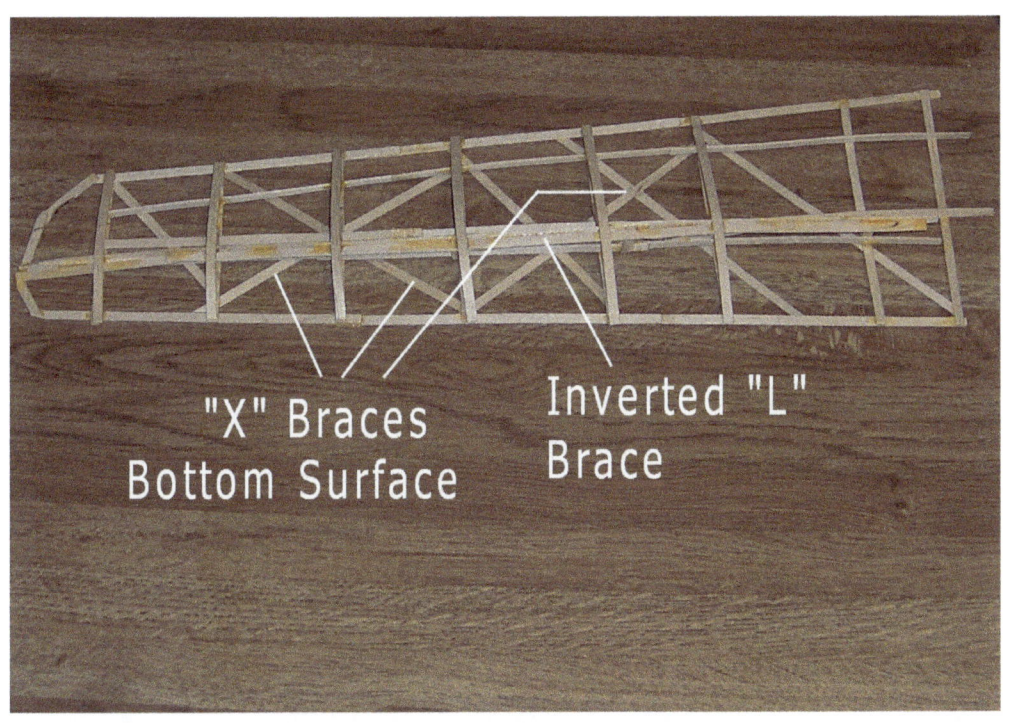

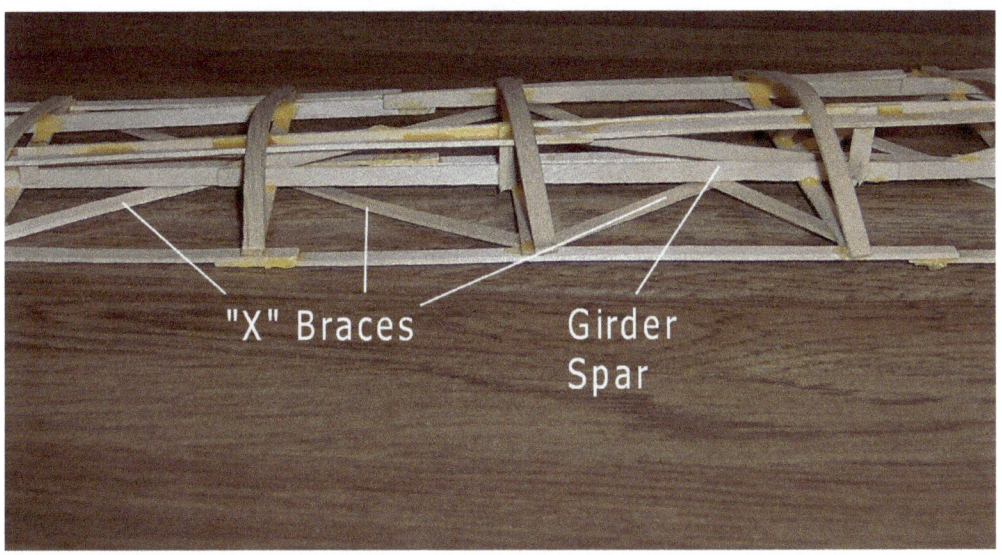

Figs. 43, 44. Wing Bracing

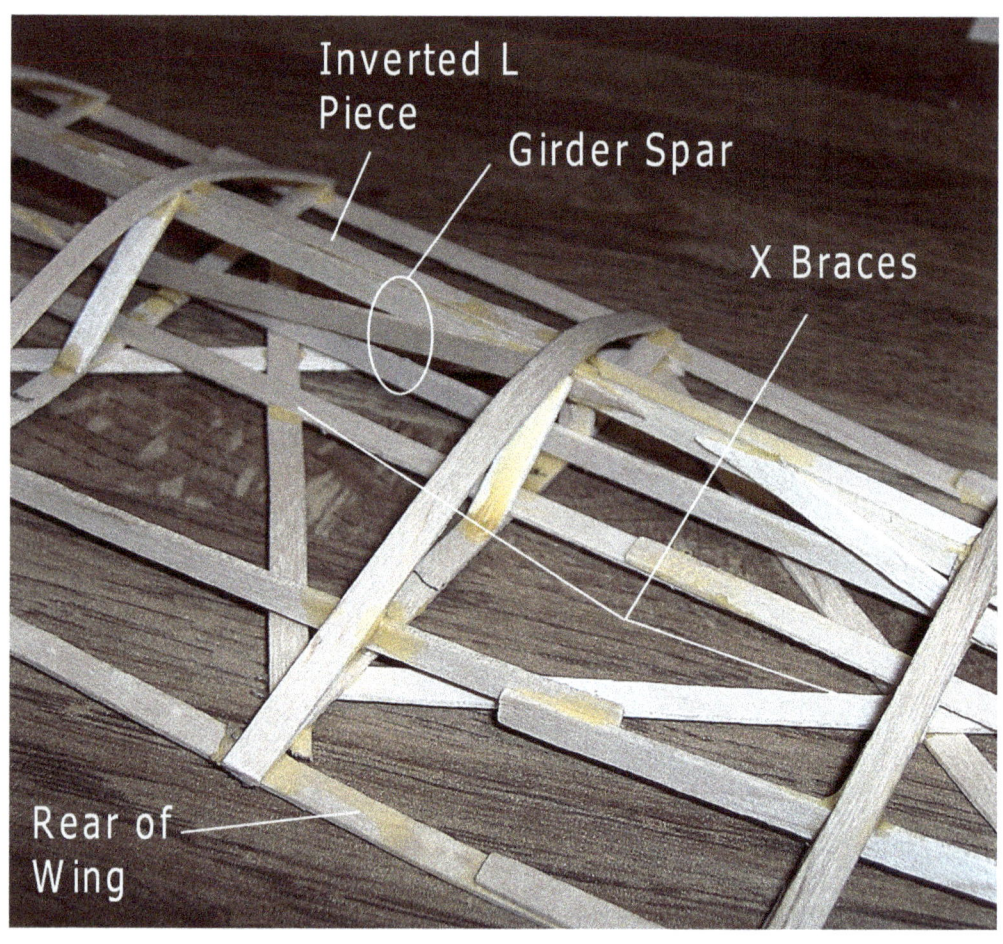

Fig. 45. Wing Detail

See the finished wing on the next page. Make a mirror image version for the other wing.

Set the two wings base to base with about the wingspan you desire, about 20 to 22 inches.

Elevate the outer end of each wing about 1/2" to 3/4", this is like the slope, called *dihedral*, that actual wings have. Now, add some joining spars at the center to form the whole wing as the P-39 has a wing right at the bottom of the fuselage. Allow a minimum for now before attachment to the fuselage when you

can reinforce it more.

Fabricating the Fuselage

Next is the fuselage construction. As you examine a photo of your particular model, you can copy its shape as you build the fuselage. Some planes had a body shape joining the cockpit portion, like the P-40 Warhawk whereas others had a bubble cockpit sticking above the fuselage like the later P-51 Mustang and P-47 Thunderbolt.

In looking at the proportion of your wings and the photos of your type plane, you can decide on a length and use a set of about 8 adjoining stirrer sticks with a 1/2" overlap as described and glued together at the beginning of this chapter.

Soak some six stir sticks in water for a bit, change ends too after about 5 or 10 minute to have the entire stick wet. Once flexible pick a stirrer and gently bend in 1/8" to 1/4" increments against your fingernail working around the hoop you are forming; the strip can easily be formed into an oval hoop this way. Probably do bends of 45 degrees as you move around the length. Then applying some glue, overlap a 1/2" area and clamp with a hobby clamp to form a hoop. Shape it a bit to be in alignment and of an oval shape.

Make another hoop a bit larger using about 1 stirrer plus ¼ of another wet stirrer... use two clamps to attach while the glue dries. Shape it onto an oval similar but larger than your first hoop. Allow to dry; these hoops will be very sturdy, and will provide good support as you fabricate the fuselage.

The hoops will be initial bulkheads in shaping the larger portions of the fuselage. Start at the rear, and use your picked airplane to set up the very rear of the tail first.

It does not need to be in a point as mine was; you can use vertical pieces of stir sticks to set the height at the rear and a horizontal crosspiece to set the width. See Figures 47 and 48.

Doing a "cross" shaped piece allows setting the height and width of narrow parts of the fuselage.

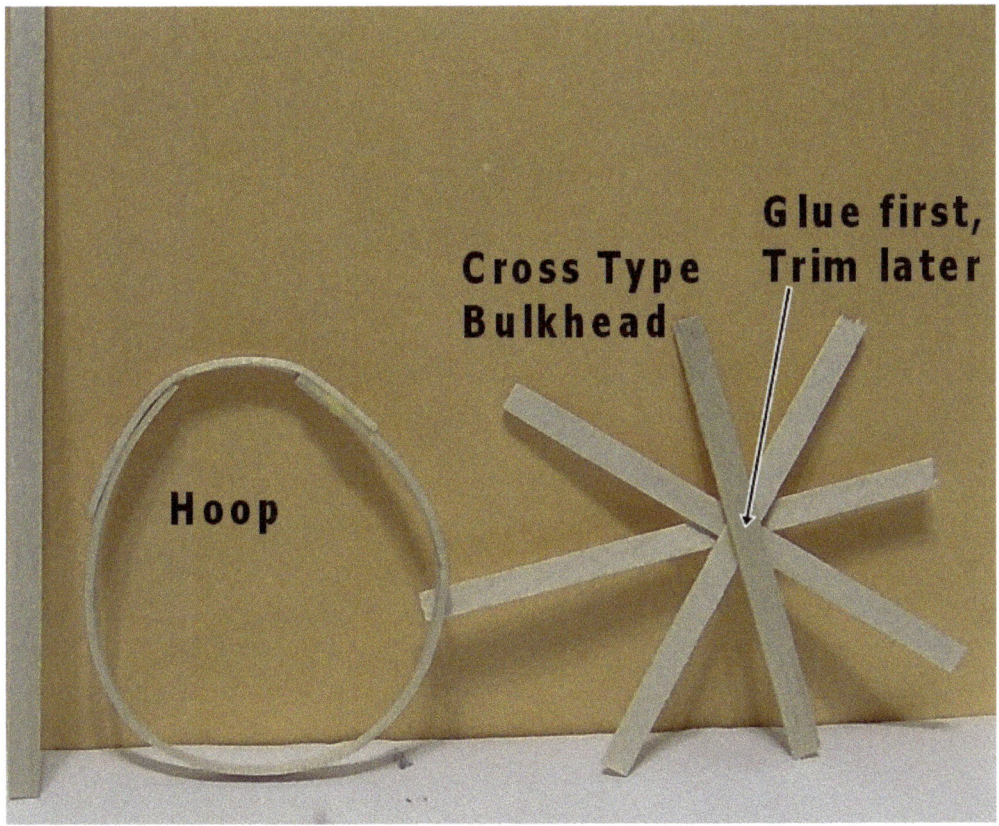

Fig. 47. Simple Bulkheads

A good way to start is with four lengthwise spars to give a good beginning towards the contour of your plane. Later you can add additional long spars to round out the fuselage.

Rubber bands are good for holding the spars in place on a "cross" style bulkhead, but the small hobby clamps are excellent for the hoop glue points. As a hoop or two are clamped, along with the cross type bulkhead, you will see a nice method for working progressively towards the fuselage shape. You can cut out or leave gaps to allow for wing or cockpit placement.

Be patient if using carpenter wood glues as they may take most of an hour to dry, go research some other scientific projects or gather more materials for a project, or start some other parts fabrication.

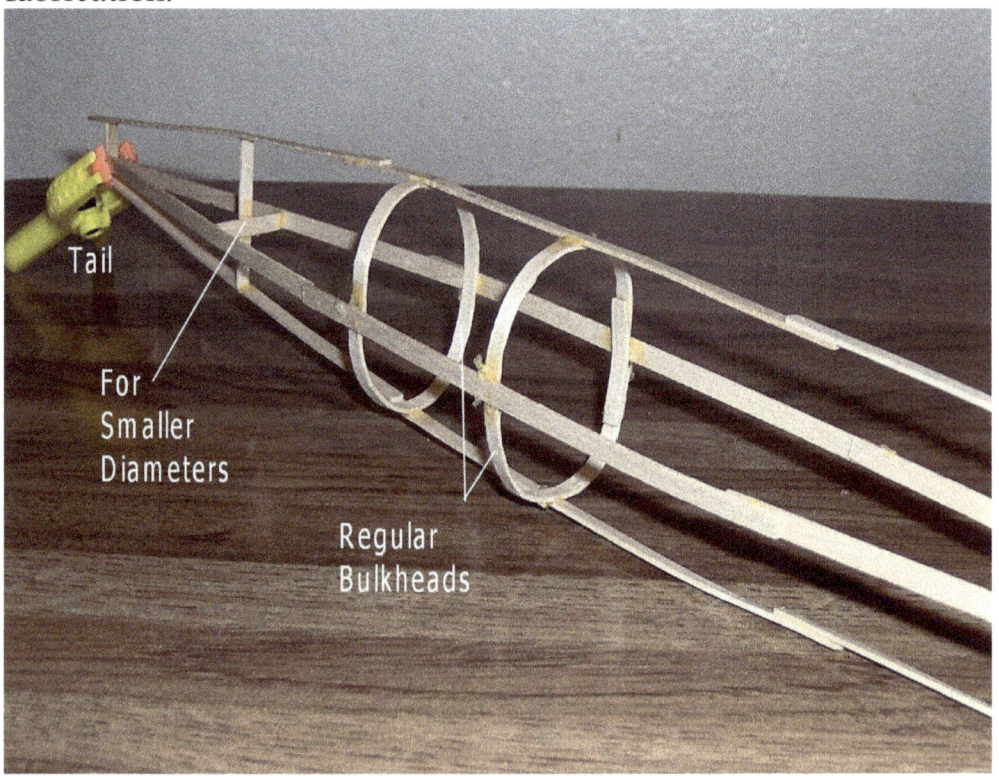

Fig. 48. Beginning Fuselage

Once you have a shape similar to your desired model using the varying bulkheads with four long spars, you can add additional long spars in between each of the four to start to get the round shape more closely. You will notice how strong those bulkheads are and for the crossed types, you can add more spokes to help strengthen the new spars. Determine where the new spar will go, then add a "spoke" to the small bulkhead, clamp it to the existing "cross" piece and trim it later after it has dried. **Do not add the** *new spars until you have added the spokes, otherwise you will not have enough space for clamping.* Sometimes it will be tricky, you can then use electronic clip lead clamps. You can use a small pair of electronic wire cutters to trim the support spokes. See Figure 49.

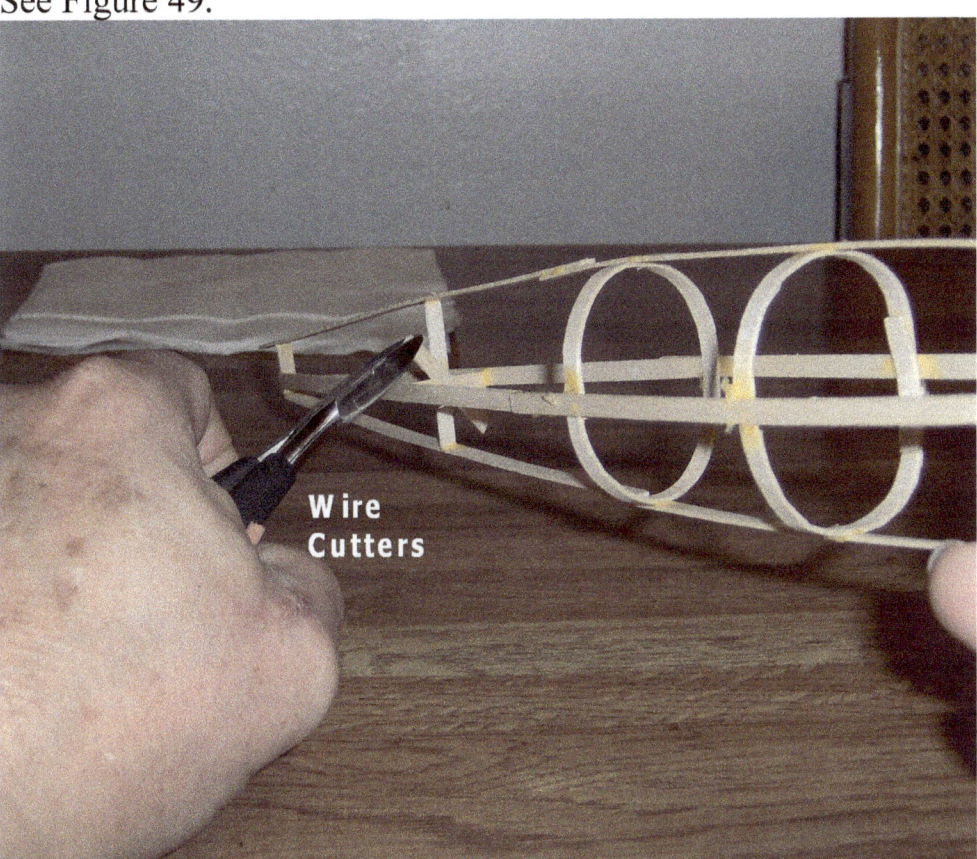

Fig. 49. Trimming Bulkhead Spokes

   Keep in mind the area for the wings and for the cockpit as well as the style as you continue with the assembly.
   Mount some vertical spars to develop the Rudder portion of the tail. You can wet and bend the sticks for curves. You can also do the rest of the tail portion like miniature wings. Figure 50 shows additional fuselage spars and beginning of the Rudder.

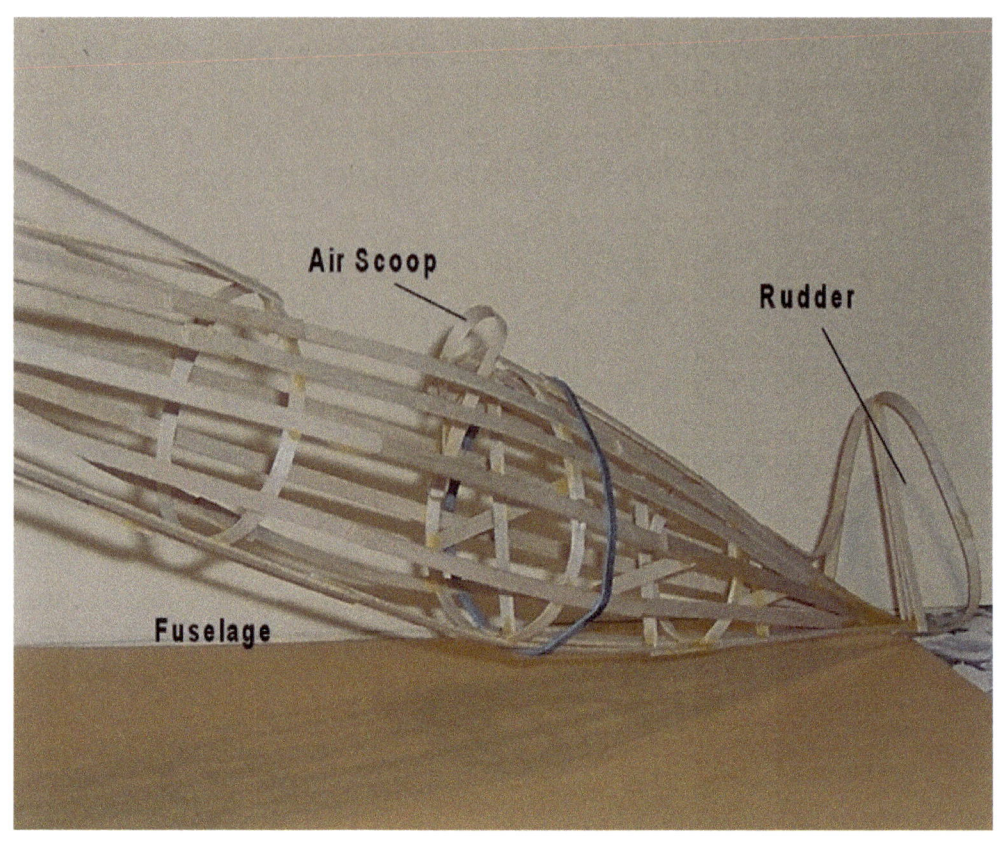

Fig. 50. Starting Rudder and Scoop

As you continue with the fuselage, try the wings in the proper area to see how they fit. If necessary, adjust or cut out a spar to check the fit. Once you have developed an opening for the wing to fit through, you can add strengthening braces as desired to make the fuselage-wing connection more sturdy. Additional bulkheads or even partial curved pieces can be used to brace and strengthen the wing to fuselage attachment. Along with setting the wings to join into the fuselage, you also want the two wings solid to each other now. By now you have learned the strength of triangles and bulkheads and should be able to do this as you study the pictures over the next few pages..

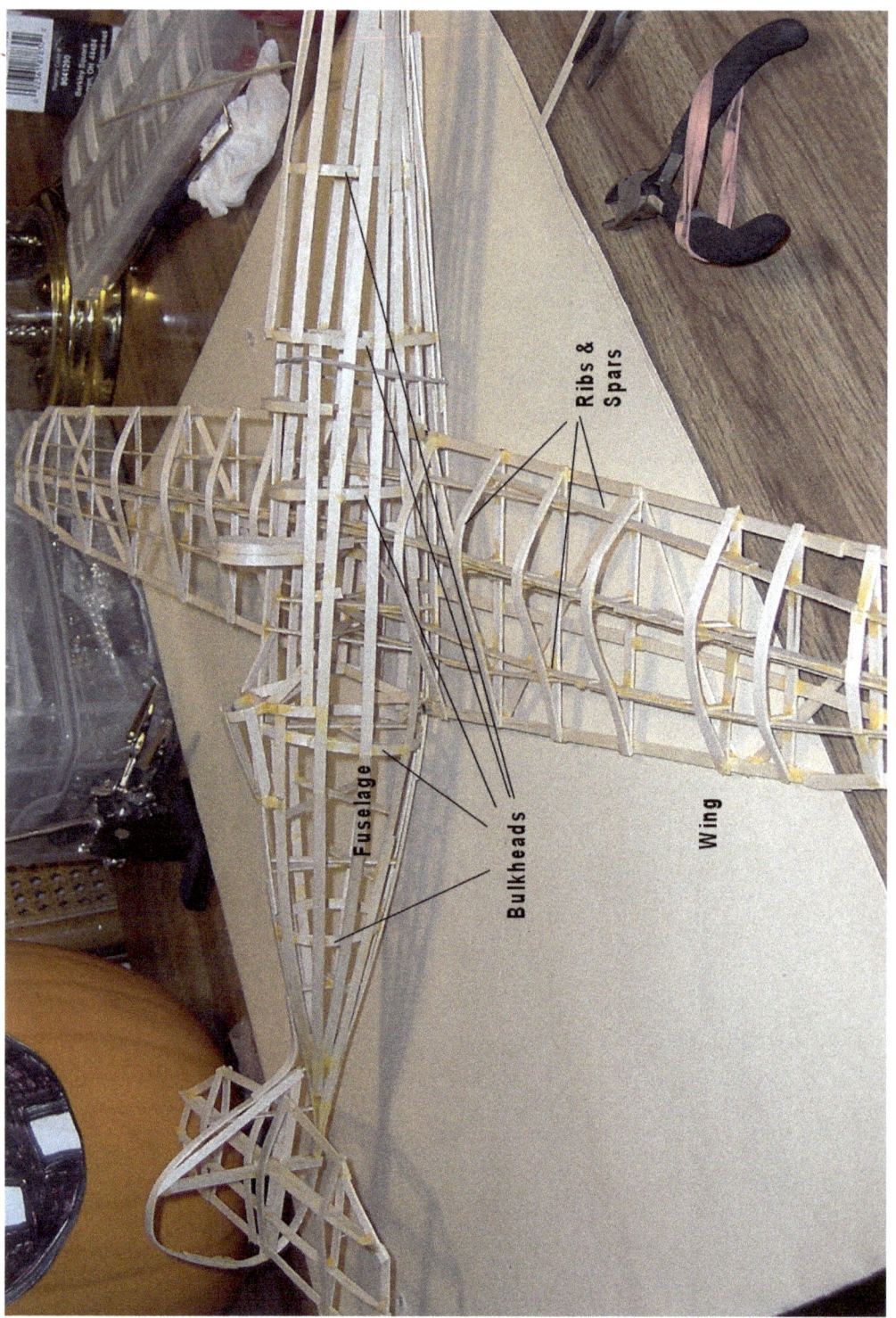

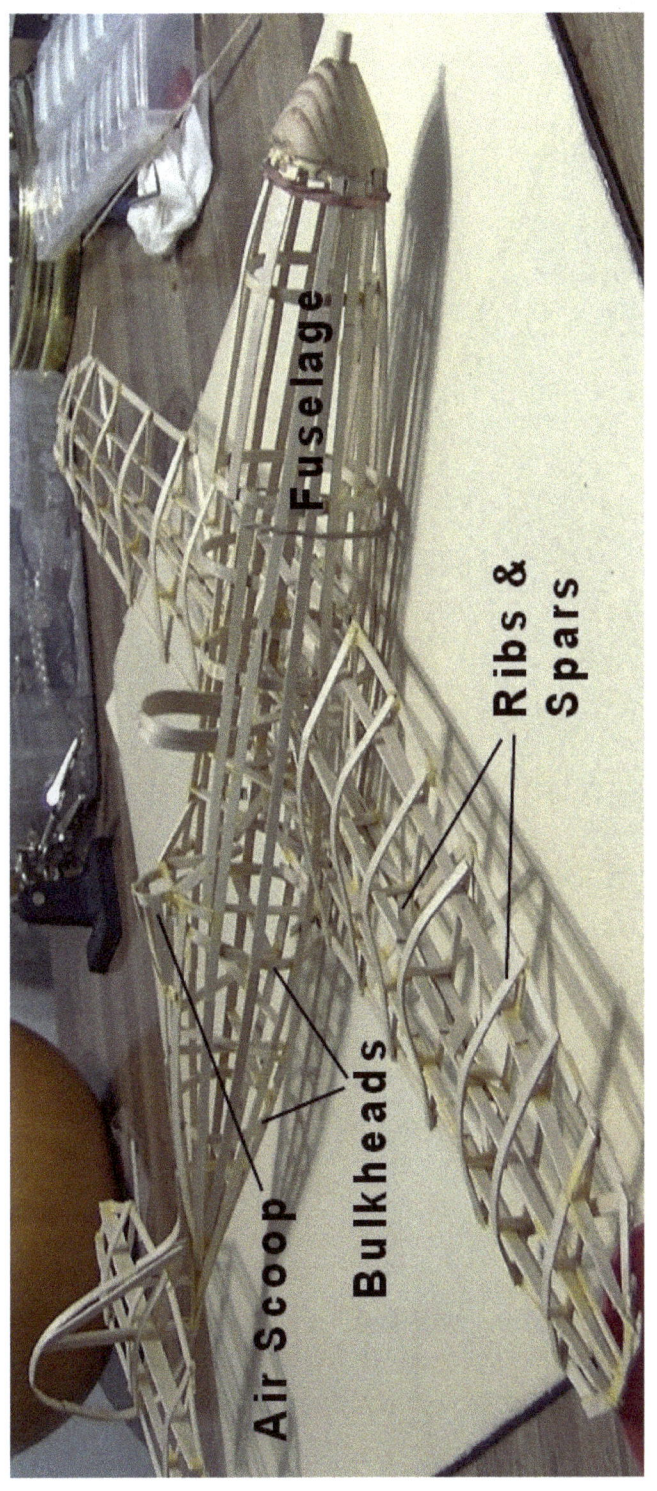

Fig. 56. Starting Skin

The previous group of Figures have shown continuing construction of the P-39 Airacobra. As you add length to long spars and bulkheads. The aircraft is taking shape! Try to match it somewhat to the model you picked. Also add in a bit of framework for your horizontal tail portions, the horizontal stabilizer.

Fig 57. Continuing the Covering

As you near the front of the fuselage, start doing smaller bulkhead hoops. You can make a small circle of thin wood to be the front piece to glue the prop hub onto. Make hoops or bulkheads to taper the front of the fuselage to be similar to your planned model.

Once done with the framework, it is time to begin covering the surface with grocery sack paper. Apply glue along a few spars and cut and form the paper tightly as possible around the fuselage and clamp to hold until the paper has grabbed. Work gradually around the fuselage. If you apply glue at the underneath edge of the top piece of paper as you form it over the underneath paper to join the two, you can smooth the surface to wipe out excess glue and smooth it together to make it stick.

Once done with the fuselage, do the wings in a similar manner, applying glue to each airfoil spar, bringing the excess to the rear of the wing after folding and glue it together.

Fig. 58. Next Portion.

Shape a small block in the end of a wood piece into a propeller hub; cut rough with a Coping Saw, then round and smooth with the Belt Sander. Do not cut it from the piece until you have it nicely shaped. The full piece of wood gives you a safe handle for finishing the propeller hub. Once satisfied you can cut it off.

Fig. 59. Complete Covering

The next thing is doing the cockpit portion. Note the canopy framing and use of a portion of hoop. The Airacobra had actual doors on each side, so the door frames are done with Baling Wire at the top window portion and the front of the canopy to simulate the thin frame around the glass.

I found a bubble pack from a glue ("Goop") which was perfect for doing the canopy, **but before doing the canopy,** you should spray paint the airplane with a suitable color. Some typical warplanes are in camouflage while others are in subdued grays, greens, or similar colors. You must do the spray coat before you mount a cockpit. Apply some latex caulk to the joint between the wings and fuselage and smooth with your finger to make a nice joint. Caulk even comes in gray for a good color match.

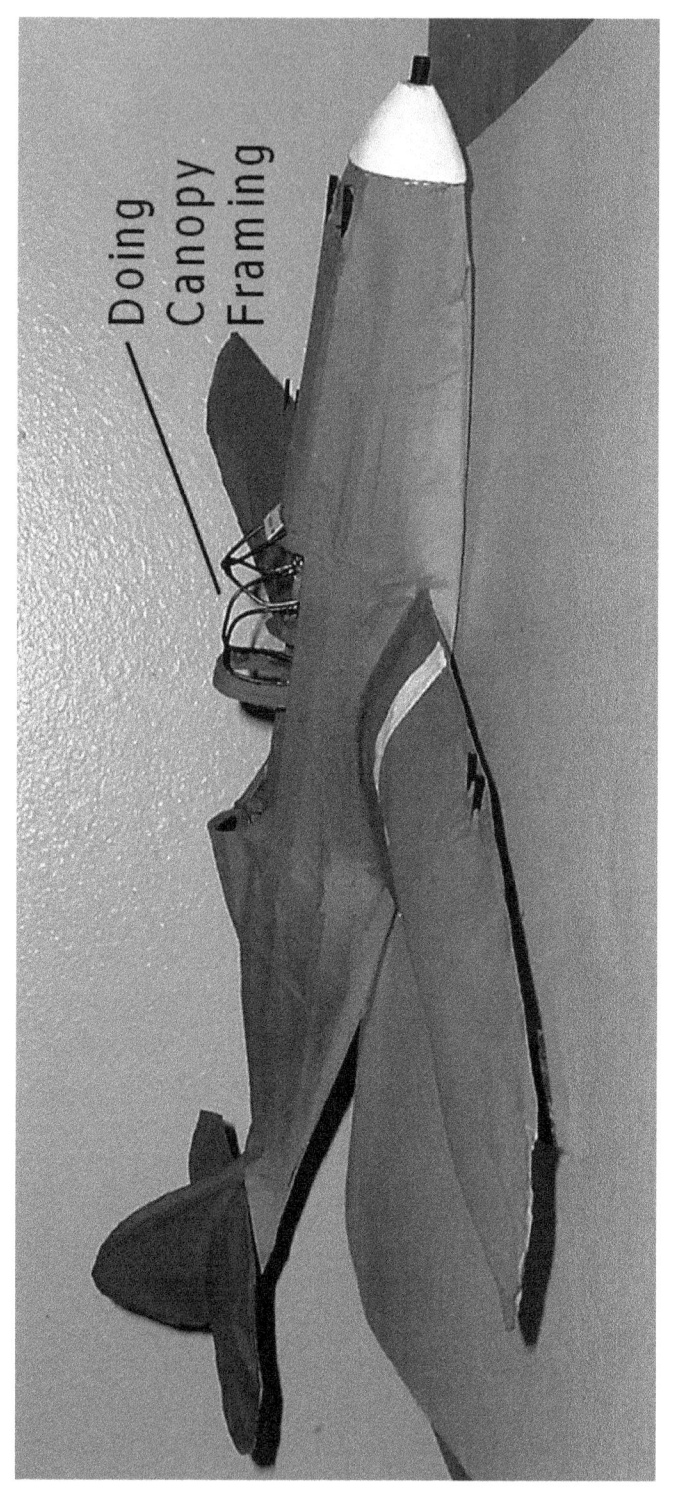

Fig. 61. Detail Doing Canopy

The cockpit is the last portion of fabrication on the model. After painting, find some bubble packs which suit your model if it has a bubble canopy. Sometimes you may have to improvise a clamp for these odd shaped jobs; note the Baling Wire clamp used in Figure 61. Use a glue which can handle plastic or wood for this job. I used Loctite Handyman Glue to glue mine onto the paper body.

You can pierce the paper skin to glue the gun sticks into place.

Fig. 62. Wings Moved Down

I realized the P-39 had wings right at the bottom, so I redid them in my model to be right at the bottom of the fuselage, not partway up. This required redoing the skin a bit. But comparing this to the actual photo at the front of this chapter shows it is more correct. Figure 62 shows after the wing was moved.

The final airplane after painting is show in the last two photos. It looks fairly close to the real photo of the P-39. The P-39 was used in so many countries, you can come up with many color schemes from the Internet.

Chapter 6

Pull Toy Dog
For a Small Child-
Two Versions

Items needed: Wood, 2 x 2" nominal,
scrap of 3/4" thick wood,
1/4" dowel for axles,
1/2" thick wood, (drawer side material..)

Simple Dog

It is fun to make a small toy for your little brother or a small child; and this is something really nice to do for another.

The toy shown in this chapter is a pull toy dog, with wheels and a string attached so a child can pull the toy as they learn to walk, and it will encourage them to hold onto the dowel attached to the string as well as go faster as they learn to walk and run.

The neat feeling of holding the wooden dowel is probably a good training for grabbing and enjoying shapes for a toddler... perhaps later holding Dad's finger when crossing the street.

The wheels are made of 1/2" thick pine or alder material; the same material that is ideal for drawer sides. Using a 1 3/8" Hole Saw makes nice wheels for a small dog as described here. If you wish a larger dog, scale the wheels upward proportionally. For twice as big, make the wheels about twice as large.

When making these wheels, be sure to clamp the workpiece securely and insure that the saw will not be over the top surface of your table and cutting into your worktable... keep the area you are drilling cutting clear of all surfaces. Take care to drill straight into the piece so the center hole for the axle is not slanted.

Wear your goggles! See Figure 65 and 66 on the next page to see the pieces required for our little project.

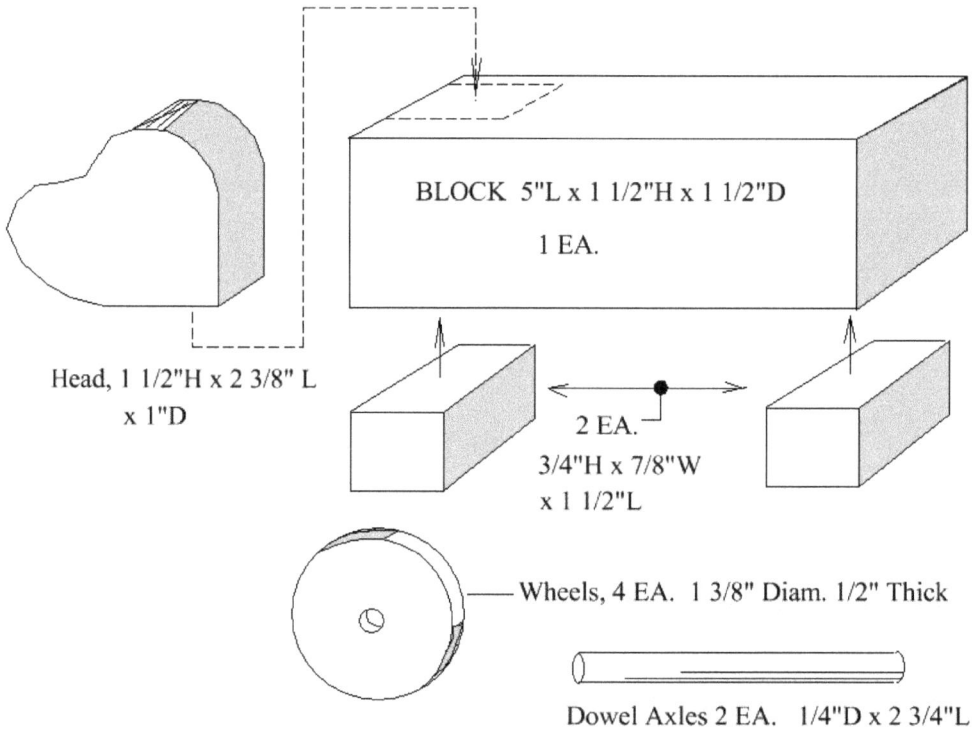

Fig 65. Basic Parts for Pull Toy

These pieces can be of any convenient wood, and once cut they should be sanded nicely to give a clean appearance. Put on your goggles and face mask. Use a Belt Sander to round all edges so no sharp edges remain for a child to be hurt.

The lower rear of the head should remain flat to glue onto the body. The top rear of the body may be cut a bit at 45 degrees and sanded to give a very round edge.

Next, apply glue and clamp the two axle blocks crossways to the front and rear bottom of the body piece. This should give an appearance as in Figure 66.

Fig. 66. Main Pieces

The next part to be glued on will be the head. To insure a good secure bond, the head should be **doweled** to the body. Dowel joints provide a large glue surface to provide great strength to a joint connection. Care must be taken to match holes on the two pieces for straight alignment and a flat joining surface between.

Fig. 67. All Parts

There is a nice tool to make this correct alignment possible, consisting of inserts with centering pinpoints … one surface is drilled for the dowels, then the inserts are placed. The other piece is placed correctly and pressed onto the pinpoints to mark it for the matching holes. Once drilled, the edge of each hole is slightly beveled with a countersink to allow escape for excess glue.
HINT: When marking the drill points, use a center punch or Awl to make them more visible, then use first a smaller bit to do a good marking hole. Once that is done, use a 1/4" bit to allow for

1/4" dowels and drill carefully perpendicular into the two surfaces. See Figure 68.

Fig. 68. Doweling Illustrated

After drilling, check the depth with the Awl and cut two dowels to fit. Be certain the length will be enough to join the head to the body, but not be so long as to block the full mating of the two pieces. Once certain of the fit, apply glue to each dowel and the mating surface and mount the head onto the body. Clamp securely, wipe off excess glue with a damp paper towel and allow the glue to dry.

Next comes the axle setup. In drilling these holes through the two small blocks beneath the body, take great care to keep them true. Perhaps a drill press is available and a Dad's help.

It might be a good idea to drill a carefully centered 1/8" hole from each side of the block if a drill press is not available. Then use perhaps a size just under a 5/16" bit as the final size to allow clearance for the 1/4" dowel axles. Center the hole in each block, and drill carefully.

Fig. 69. Finished Ready for Final Painting

Use two 1/4" dowels about 2 9/16" long for each axle. Apply glue and mount a wheel on each one, make sure the dowel is flush with the outer wheel surface. Allow to dry.

Once dry, feed a wheel assembly through each axle hole in the dog body.

Apply glue to the remaining wheels and push them onto the other end of each axle, but be careful to leave a space between the wheel and dog body to allow easy rotation of the wheel. There should be about 1/8" of space. A 2 9/16" axle length should be close when the wheels are pushed on with the axle flush to the outer surface. Remove any excess.

Drill a 1/4" hole in the rear of the dog body. Later you can knot a piece of heavy cotton cord to make a knot which can be shoved tightly into the hole after applying glue. Then you can cut the end off to simulate a little tail. ***Wait until the dog is finished to add the tail.***

Next, drill a larger hole about 3/4" deep into the front of the body to accommodate a 7/16 to 1/2" diameter dowel. See Figure 68 which shows the front dowel with a small through hole to allow tying a pull string onto the toy.

Do not add the string until the dog is varnished or a Danish Oil finish is applied. In a toy like this, with rubbing parts, Danish Oil is a best choice. Sand the front dowel to allow a fit if needed, then apply glue and tap the dowel into the hole.

Drill a centered hole into another 2 ½ to 3" long piece of the large dowel. This will later be tied to the other end of your cord. This is the pull handle.

Paint or use a Sharpie Pen to draw the face onto the dog. Remove any axle stub protruding from any wheel and sand to finish.

The entire dog can then be covered with natural Danish Oil to give a nice finish. Follow the manufacturer's directions in applying the finish. Be sure to place the damp application cloths or paper towels into a waste can that is dumped quickly. These items could spontaneously catch fire if not disposed of correctly.

Once all finished and dry the cord and tail may be added.

The finished dog is shown in Figure 70.

Fig. 70. Finished Dog

Next-- a Segmented Dog

   The next dog shown is a bit more work as it uses small blocks strung together to form the body; it is a little version of a dachshund (Wiener Dog) and the little blocks turn every which way as a child pulls or plays with it. The body is flexible and sags somewhat. Just like a real Dachshund.

The one shown here uses seven blocks of approximately 3/4" thick material that are about 1 1/2" x 1 3/4" across. See Figure 71.

Drill a 1/8" or so center hole through each of the seven blocks in the center as shown in the picture, carefully aligning them. You can drill two at once then use one as a guide to start the hole into another block. When all seven are drilled, they should look like the ones in Figure 71. Later a cord will string the blocks together to form the dog body.

Sand all corners and edges, and round the rear block a bit extra.

Fig. 71. Drilling Blocks

The front and rear block will need to be drilled for wheels, very much as was done in the previous dog construction. Drill through the blocks crosswise, carefully using a 1/8" bit initially. Then use a bit almost 5/16" as the final bit to allow for easy turning of 1/4" axles.

Fig. 72. Head and First Block Assembly

Cut four wheels using the 1 3/8" diameter Hole Saw, just as used for the previous dog. Sand around the edges to smooth them.

Cut out a dog head shape somewhat as described in the last dog project, Round sharp edges, and dowel it into the front block for a strong joint. There should be glue applied on the flat surface as well as the dowel assembly. Push the head and front block together. Wipe off excess glue with a damp paper towel. Clamp the piece for a good joint. You will now have a front block and head unit as shown in Figure 72.

Cut two 1/4" dowel axles to about 2 3/4" length. Apply glue and mount a wheel onto each one so the dowel is flush with the outside of the wheel. Wipe off excess glue. The parts now look as in Figure 73.

Fig. 73. Parts For Dachshund

Draw the eyes, nose and ears onto the head using a Sharpie and inspect all sanding, sand as needed.
Apply glue and mount the wheels onto the front and rear blocks of the body. Allow about 1/8" total clearance between the wheels

and blocks. After the glue has dried, cut off any excess axle beyond the outer wheel surface.

Make a small length of 1/2" dowel as the pull handle and drill the center for the cord.

Once again, with the moving parts, I prefer Danish Oil; it is clean and smooth, doesn't stick parts together. Apply the oil to each piece, wait as instructed on the can, then wipe off. Perhaps a second application can be used. If you wish, you can do a bit of fine sanding wet with 400 grit on this second coat; the parts will look beautiful. Then do a final wipe and let dry overnight.

Final assembly consists of pushing the cord through each piece starting at the tail, and after coming out the last front block, add maybe 4 feet excess. A trick to make it easier to thread the cord through--- apply some glue to the end of the cord, and after it dries, nip it to a point.

Tie a multiple knot at the front to lock the front block and tie the end through the hole in the pull handle.

At the rear of the dachshund, snug up the blocks, and do a multiple knot to lock them together snugly. Then leave maybe 2" of string frayed as the tail.

I used Golden Oak Danish Oil for the finish, and the final toy is shown in Figure 74.

Fig. 74. Finished Dachshund

Chapter 7

Building a
Real Compass
Yourself and
Why We Use A.C. Power

Items needed: 3/4" thick wood Piece,
Sewing straight pin,
1/4" thick plywood panel,
1/8" thick Plexiglass panel,
Brass screw and washer

Although you can buy a compass at the store, it is interesting and fun to make your own, and it is not difficult. The compass described here is made into a small case made of a scrap of wood. The needle is made from a piece of tin can. Tin cans are actually Steel with a thin Tin coating.

There is a book, "The Dangerous Book for Boys", which claims you can rub a needle with silk and make it into a compass needle. Also, the Alaskan survival movie, "The Edge" claims the same thing. Those who came up with this idea apparently mixed up the property of Static Electricity with Magnetism. These are two different Scientific Principles.

Rubbing a rubber or plastic rod with Silk or Cat's Fur can impart a static charge which will attract bits of paper, (We learned that in Science years ago...) but it has **no connection**

with magnetism, or magnetizing a needle or steel object.

When being magnetized with D.C. Voltage, a needle would be typically placed inside a coil and a momentary D.C. Current would be applied. The needle will always magnetize the same assuming the same polarity and needle placement, and same polarity on the coil..

However , electricity present in every home and factory does have varying magnetic fields present in any wiring while passing electricity. These fields **can** impart magnetism.

Why does A.C. Power cause a needle to magnetize?

It is now time to discuss A.C. power a bit. When you turn a device on or off, the magnetic field emanates out from the powered device, and the field is either building or collapsing constantly.

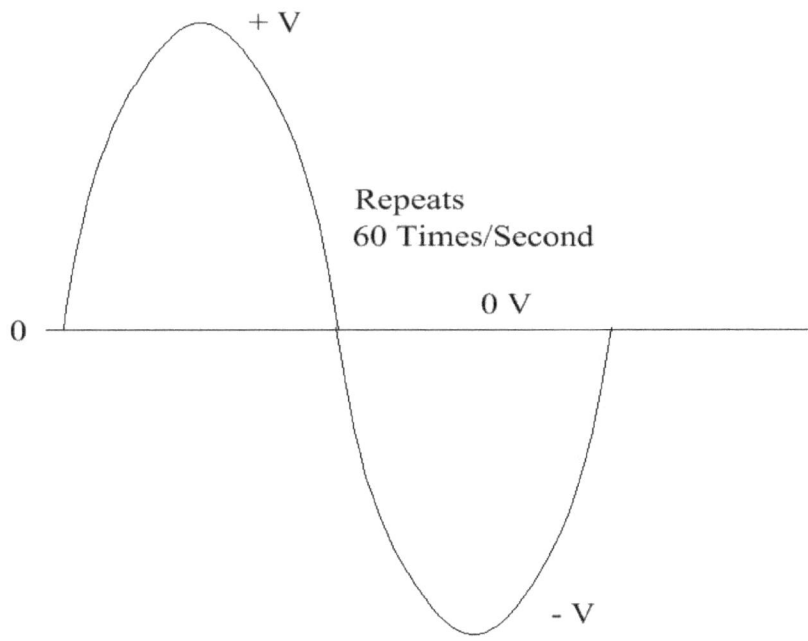

Fig. 75. A.C. Waveform

A.C. Voltage in the U.S. varies from a positive value down through zero to a negative value 60 times per second. Any device powered by A.C. (Alternating Current...) like refrigerators, sewing machines, blenders, motors, etc., basically all units powered by electricity, can cause a magnetic field to effect a nearby needle or metal which can be magnetized. Every wire has a magnetic field around it. Thus any needle laying around the house could already be magnetized.

   It is actually quite likely, for example, a sewing machine gets powered on and off and therefore a nearby sewing needle would likely be magnetized from electric fields.

   Normally the upswings and downswings in the A.C. Voltage would actually ***demagnetize*** a metal object as the direction of current constantly changes back and forth. It takes a current ***in one particular direction*** (D.C. Electricity, like in a battery...) to magnetize and form North and South poles in a metal needle or other steel object.

   So, how does A.C. Power magnetize anything?? It is because during switching off, the voltage is more likely to be at a positive or negative value during the exact instant of shutoff. If you look at the waveform, only two times per cycle is the voltage exactly at zero! The collapsing Voltage and electromagnetic field thus is usually at some value as it shuts off, so the magnetic field around the wire, or motor wiring collapses from one particular direction.

   Imagine a needle laying near a sewing machine. As the sewing machine is turned off the electromagnetic field enveloping the nearby needle collapses from some value and it acts upon the needle, either magnetizing it in one direction or the other... depending on which direction the voltage is collapsing from and how the needle is laying. A needle will have a North end and a South end, if magnetic. With A.C. the needle may magnetize one way one time and the opposite the next. But it is very likely to be magnetized!

And why do we use A.C. Power?

Back in the time of the inventors, Thomas Edison and Nikola Tesla, around the early 1900's, there was intense effort being made towards running power over a distance.

In running power through wires, there is a loss because of wire resistance... any metal conductor has some property resisting the current flow through it. There is a certain quantity of voltage force required to push current through the wire. The loss of Voltage due to Resistance is larger and larger as the length of wire increases and the current becomes larger. This loss thus varies **directly** as the current and resistance.

Now, think about the long electric wires to power a home or business far from the power plant, even hundreds of miles away! Because of the distance and long wires, the resistance will be large causing a huge loss of voltage over the distance of the line.

The voltage loss is V and there is an equation to show the loss over the line.

A common term in Electricity is Voltage, V, and it relates to Current, I, and Resistance, R, by a formula:

$$V = I \times R$$

As R gets large due to a long wire carrying current to run a home, you can see V gets large, and this is the voltage drop on the wire, soon there wouldn't be enough voltage left to run common household appliances and lights!

Edison wanted to use D.C. , Direct Current, but over a distance the voltage loss would be too much. There was no way to provide a high voltage safely for long lines in a direct current form.

However, Tesla had invented the A.C. Generator, which could be turned by water flow in a river, essentially a free source of energy! Additionally, with the A.C. voltage and current, a device called a transformer could *increase* the generator voltage to a very high value at the same time it *dropped* the current... because a transformer operates such that as voltage boosts up, the current drops down; it thus does exactly what is needed to supply power over extreme distances!

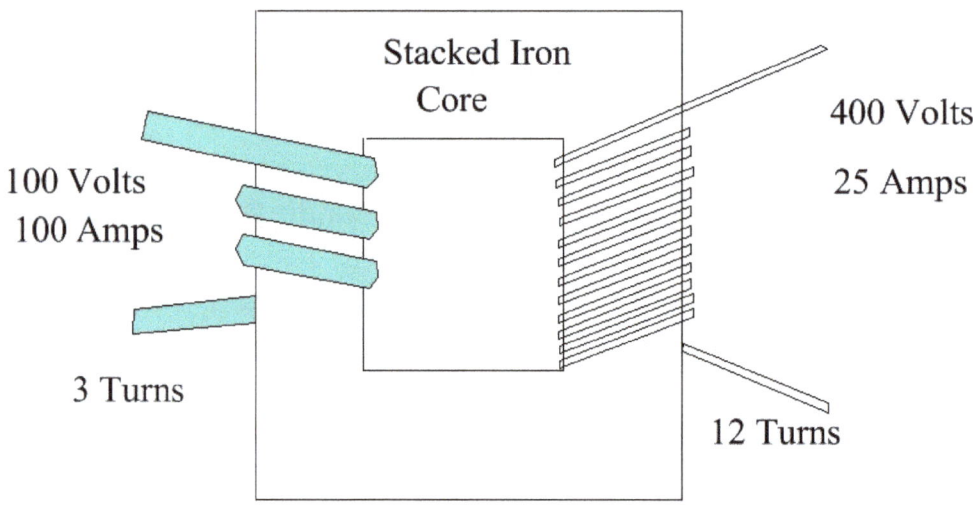

Fig. 76. Simple Transformer

A transformer is a device made of a stacked sheet Iron core with an input winding and an output winding. Voltages and currents are coupled from one winding to another due to the magnetic field in the Iron core. The ratio of turns in the two windings sets up a voltage and current ratio allowing incredible things. By using this device, voltage from the A.C. Generator can be "***stepped up***" as the current goes down proportionally to a small value. Then the equation shows that voltage drop can be small over long lines because the current can be made so tiny.

Essentially,
$$V = I \times R$$

still, but because **I** can be made so small, the voltage drop in the line is minimal. And at the receiving end, a ***Step Down*** transformer can restore the voltage and current to correct values for a home!

For all these reasons, Tesla's A.C. Power became the standard adopted all over the world!

Constructing the Compass

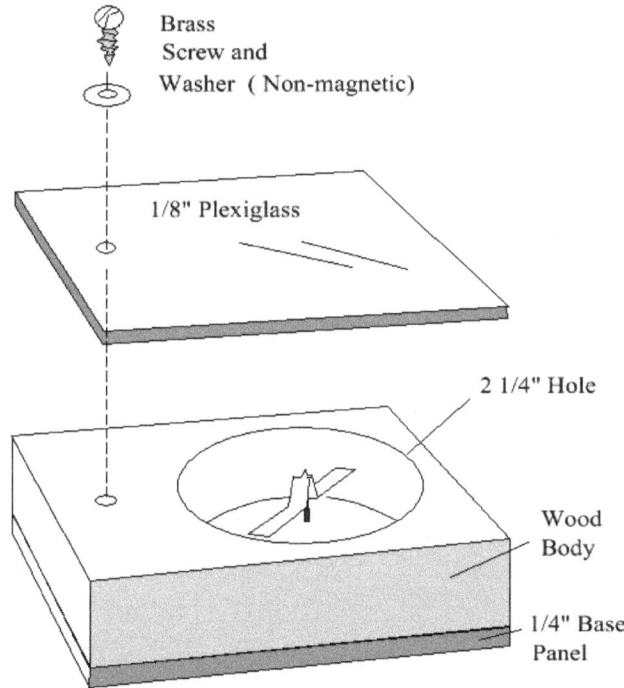

Fig. 77. Homemade Compass

You will need a scrap of wood about 2 3/4" W x 3/4" thick maybe 6" L for the body of the compass, and a piece of panel of 1/4" thickness for the base.

Fig. 78. Needle Detail

Save the lid off a standard vegetable can for the compass needle, and get a straight pin from your Mom. With your Dad's help, and a Tin Snip, cut a strip about 3/16" wide from across the can lid suitable for the compass needle. By eye, put a center punch in the center of the strip, and gently tap a small dimple into the center. ***Be careful not to puncture the tin, only dimple it!***

This dimple will form a pivot for a pin to support the needle. Next fold an inverted bend directly on either side of the dimple to give an appearance like Figure 78 . This stabilizes the needle as the weight of the needle will be below the support point.

Fig. 79. 1/4" Panel Base with Pin

Use a tiny drill or brad to punch a small pinhole into your 1/4" panel about 2" in from both sides of a corner. Wear safety glasses, and cut a small sewing pin into a 3/4" length with the point using tin snips. Mount this through the hole so it sits 1/2" above the top surface.

Fig. 80. Main Parts Finished

Put a bit of acrylic, silicon or epoxy glue into the hole to mount the needle perfectly vertical from the panel surface, see Fig. 79. Allow to dry. Make sure it stays vertical in the hole.

Now for the compass body or case. Fasten the 2 3/4" wide wood piece in a vise or perhaps get your Dad to do this next step in a Drill Press. Center between the narrow sides and carefully use a 2 1/4" Hole Saw to drill out a nice hole. Take care to hold the drill vertical to the flat surface when drilling. Once finished, sand a bit inside to smooth the hole. You now have the main parts made as shown in Figure 80 .

Trim the ends of your needle so it fits inside the 2 1/4" diameter hole with about 1/8 to 1/4" of clearance. Also set it carefully on your base pin to check its balance and trim tiny bits to get a nice balance.

Now it is time to magnetize your can lid needle. Find a magnet ... preferable an obvious two ended one, rather than a refrigerator magnet. Gently rub the needle from the center outwards with one end of your magnet, slowly a couple dozen times. This should magnetize it nicely.

(You can also use a battery and several turn wire coil around the needle with a momentary touch to the battery... however, not everyone has a coil made up.)

Carefully place the needle onto your needle base and check that it swoops to North and South. Be sure it is away from any metals that could deflect it. Mark the end that points North with a Sharpie Pen so you will know which end is the North pointer.

Place the body with hole over the needle and pin base plate. Make sure of alignment and fit, make some marks and you can glue the body piece to the base plate.

Once glued together, you can trim the compass to a nice pocket size, I suggest something like 2 3/4" wide x 4" long with the compass portion a bit towards one end to allow for a swinging cover, perhaps of 1/8" Plexiglass. Use Danish Oil on the wood pieces prior to mounting the Plexiglass and paper dial. Then mount the Plexiglass with a Brass screw and washer, so no magnetic material is near the needle.

Fig. 81. Drawing the Dial

At this point a dial should be made. Use a Drawing Compass to draw a circle about 2 1/8" in diameter in the center of a piece of paper. Then use a Protractor to draw lines each 10 degrees around the dial. Pick the four quadrants N-S and E-W to draw a crossed line centered across the circle joining 90 degree points and label them. Now, if you wish, you can place a shorter 5 degree graduation line between each 10 degree line. (Fig. 81 and 82...)

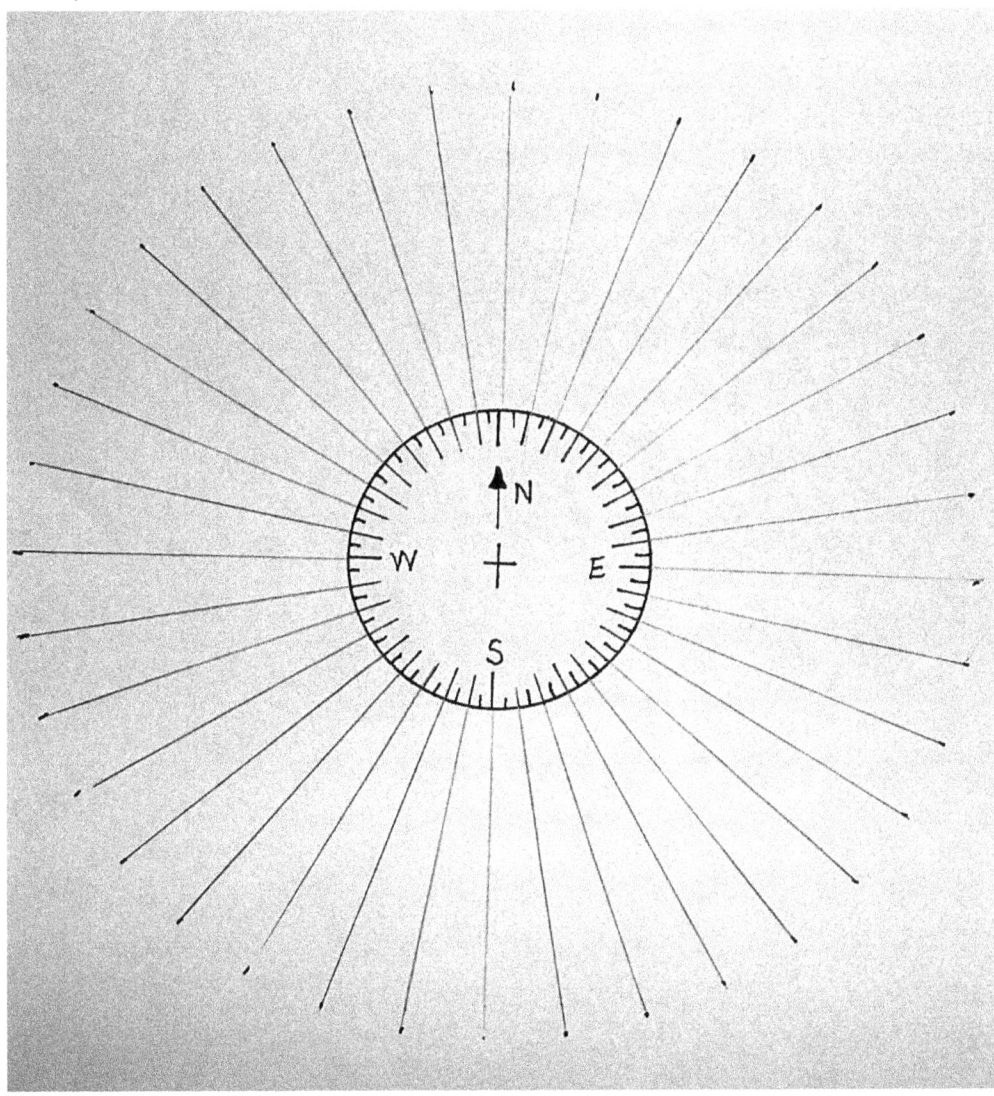

Fig. 82. Dial For Compass

Cut out and glue your dial into the compass body. Your finished compass should look something like Figure 83.
When you are not using your compass, swing the Plexiglass

cover and remove the needle and store it in the compass dial space. Out in the woods, if you wish to use it, simply place the needle onto the pin and away you go!

The compass will float but of course it is best to not let it be dropped into water. With the cover slid closed, the needle and pin are protected inside the body.

You,can round the corners of the compass if you wish, so it won't snag in your pocket.

Fig. 83. Finished Compass

Chapter 8

Making a Kaleidoscope

Items needed: Heavy Cardboard Tube,
Milky plastic cover from
an Oatmeal container,
Clear plastic,
Mirror, cut to specific size

A very neat and easy toy to make is a Kaleidoscope, and it also illustrates some scientific basis for its functioning.

Look over the parts list... the cardboard tube must be a good strong stiff tube, at least 1 3/4" in diameter. If you check at any print shop, or copying office you probably will find such tubes for free, even larger ones. (My tubes came originally from a plastic microwave wrap but now the companies appear to have gone to a smaller tube.) Also Joann's Fabrics may have some or if that fails, check at a hobby store.

Save a flat scrap of thin, stiff plastic from any packaging material; I got mine from a windshield wiper package, but many items come in such packaging.

The milky plastic can be from an Oatmeal package lid, also some other products come with a plastic clear, milky snap-on type lids, coffee, nuts and other items in can type containers.

The mirrors will come after you find your tube and do some

careful measurements.

Get a Protractor for drawing angles, and a scrap of stiff, white paper or cardboard. Lay the Protractor onto the paper and hold it steady and flat.

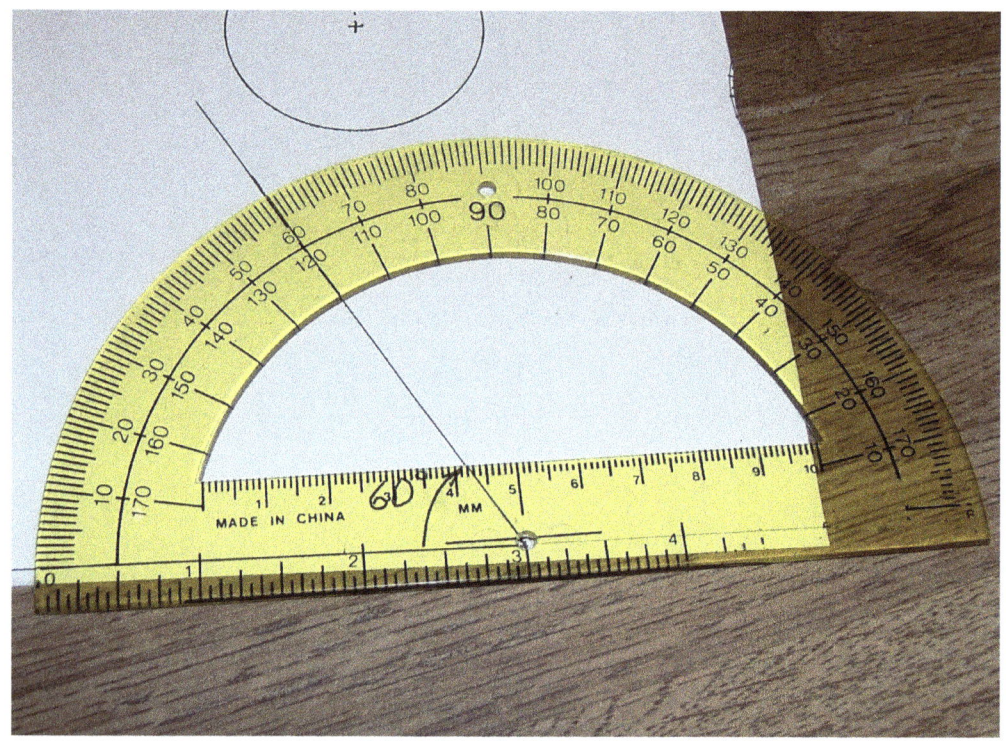

Fig. 84. Drawing the 60 Degree Angle

Mark a point with your pencil at the center point of the Protractor; then, without moving the Protractor, make a mark at 0 degrees and at 60 degrees. Remove the Protractor and draw two lines from the center point to these two points. See Figure 84.

Very carefully, cut along those lines to form an accurate pattern for determining the mirror dimensions.

Carefully place the pattern at the end of your tube with the apex (center point...) just touching the inside of the tube, and try to adjust the pattern so the two sides appear equal where they touch the inside of the tube. Measure them to get them equal.

Fig. 85. Determining Mirror Width

Once you have the two equal and just touching the inside of the tube, measure the side carefully. This sets the width of two mirrors you will have cut at a glass shop. Make the length about 3/4" less than your tube length. Then once you determine the inside measurement with your pattern, you will have your mirror dimensions. See Figure 85.

****

Subtract 1/8" from the mirror width you measure to allow room for gluing the mirrors into the tube. The mirrors will be 1/8" thick. Have them done at a glass shop.

****

Make your cardboard tube length about 4 to 5 times the tube diameter. For example if you used a 2" outside diameter tube, cut the tube length to say 8 or 9 inches. Use a Coping Saw with a fine tooth blade and take your time to cut it off straight..

In Figure 86 you can see the parts.

- The milky plastic is for the bottom cover cap.

- Above the bottom cap is a compartment to hold the small pieces that make up the pattern.

- A clear plastic portion will form the upper cover of the sealed compartment.

- Above that will be the two mirrors mounted vertically to form a 60 degree angle with each other. Each mirror is 3/4" less length than the tube length and 1/8" less than the measurement on the 60 degree pattern width side.

- At the top will be a final piece of the clear plastic and a small viewing hole.

Once you get the two mirrors, cut from the glass shop, in 1/8" thickness, wipe them carefully with some alcohol and a bit of tissue to clean them. Avoid touching the surface. Use a small bit of scotch tape to hold them together at one corner vertically. Then set your 60 degree pattern in place at the base and align the

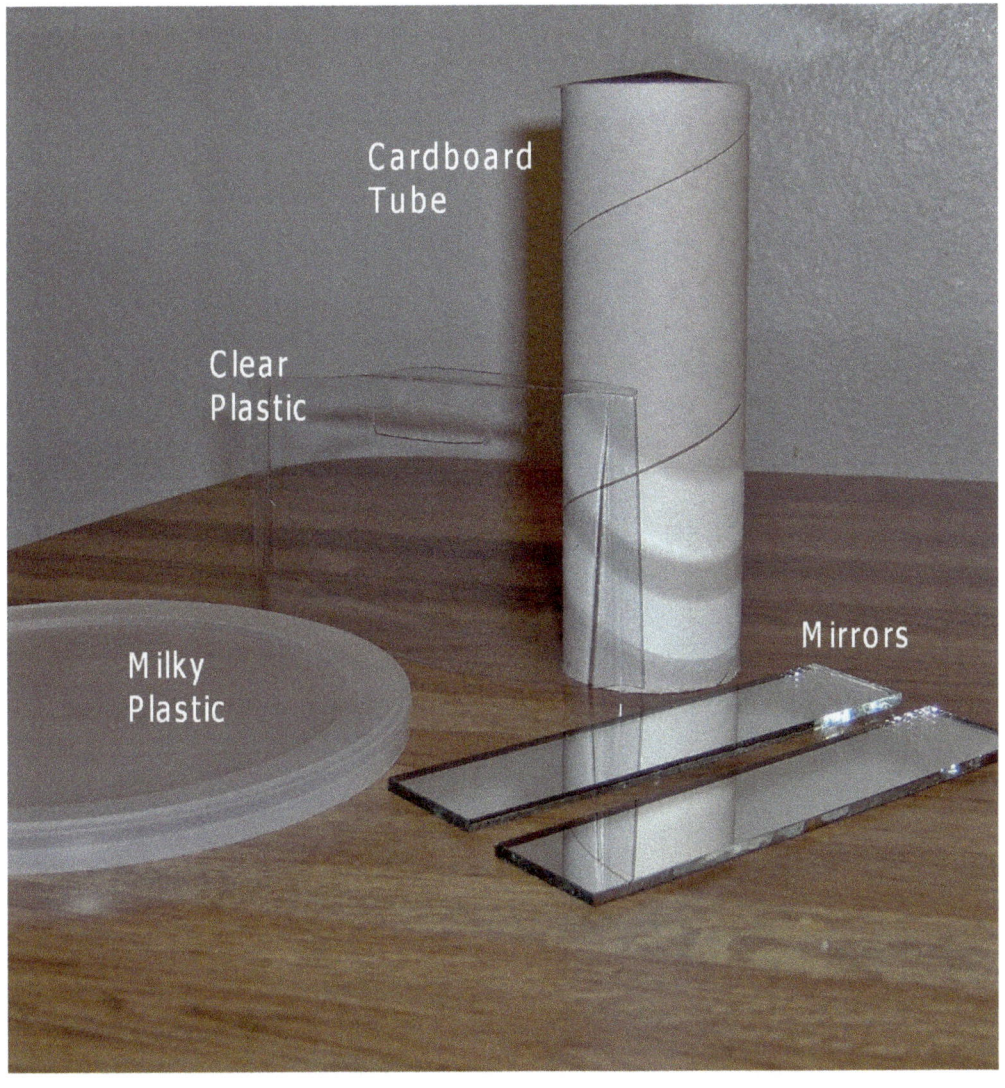

Fig. 86. Main Parts for Kaleidoscope

two mirrors to the angle. Keep the apex of the two mirrors touching. Run a bead of RTV Silicone Adhesive along the outside of the joint and allow to set. Make sure not to get any RTV on the front of the mirror pieces. The angle setup will be strengthened once you get it mounted into the tube. Be sure to allow it to dry.

Once dry, gently slide the mirror assembly into the tube. Leave the mirror assembly even to the tube at one end once you slide it in. This will be the top of the Kaleidoscope.

Fig. 87. Top Setup of Mirrors in Tube

Recheck the angle with your 60 degree pattern, and apply some Silicone RTV or Handyman's Adhesive to the corners of the mirrors at the top of the tube as shown in Figure 87. Be sure it does not protrude out of the tube, as you will also be placing a eyepiece over the end of the tube. See Figure 88, which shows the lower tube end with glue applied. Note the glued mirror assembly.

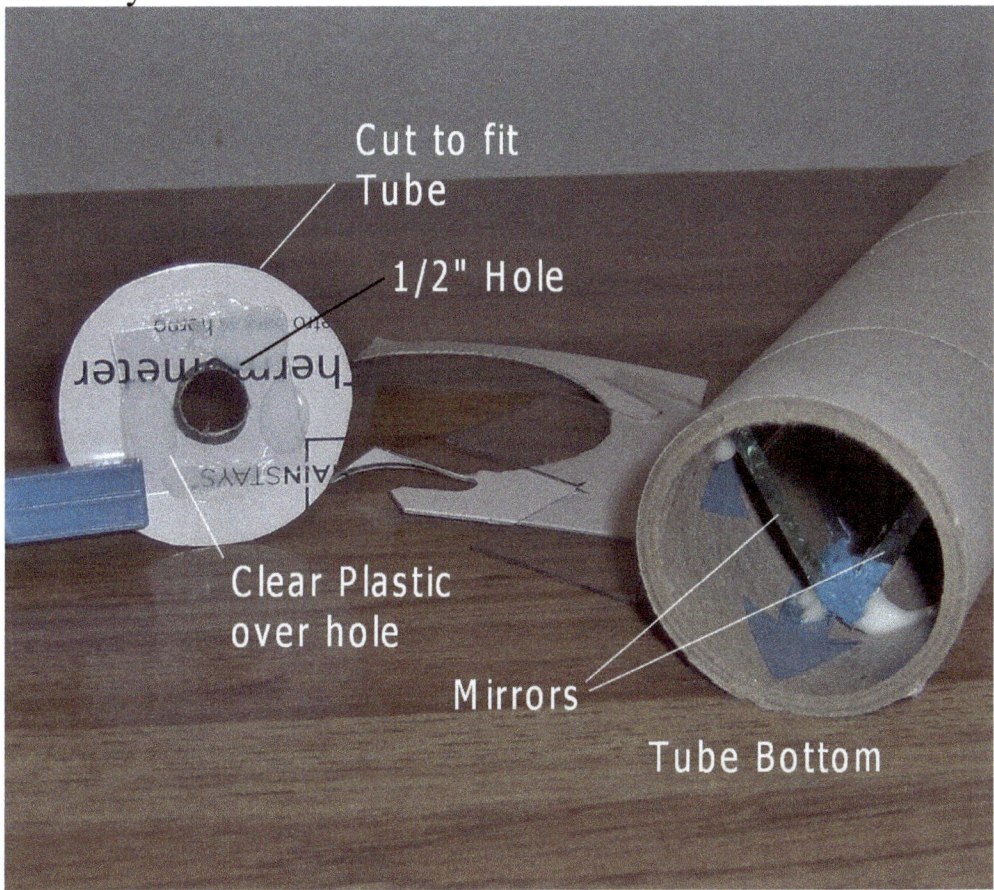

Fig. 88. Bottom of Tube and Eyepiece

Use a stir stick to apply the glue to the corners and be sure the glue does not stick out into the lower tube compartment area where the pattern pieces will go.

Cut a stiff cardboard circle (the diameter of your tube...) for the eyepiece, as shown in Figure 88. Carefully cut a 1/2" hole in the center... be careful as you will have to use a sharp point type of knife to do this. Then glue a piece of clear plastic over the hole.

Once the glue has dried at the corners of the mirrors where they touch the tube, you can use a stir stick to add some Silicone RTV down inside the back of the mirrors where the tube touches the mirror to lock them more solidly.

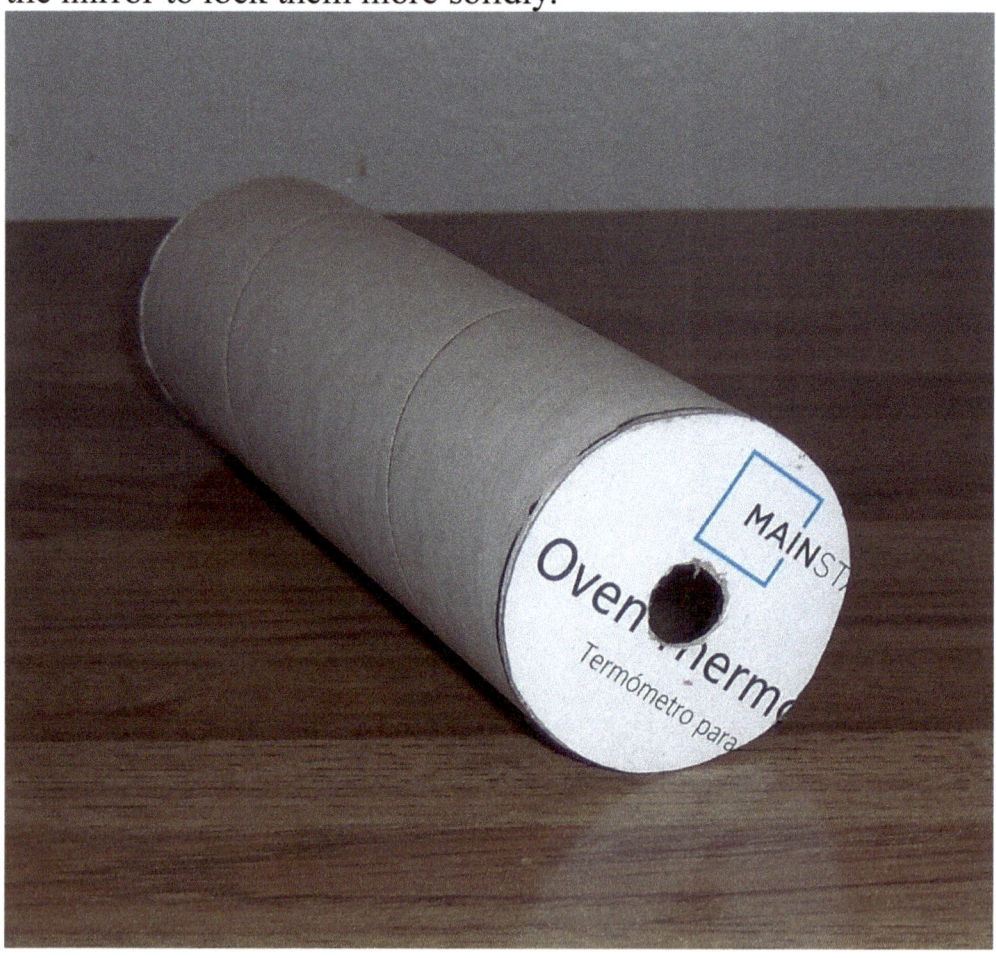

Fig. 89. Eyepiece Installed

Once all glue has dried, invert the eyepiece so the clear plastic will lie inside the tube. Glue it to the top end of the tube-- the one with the mirrors essentially even with the end of the tube.

See Figure 89.

Next it is time to make up the compartment for the display pattern pieces. Cut a round piece of the clear plastic sheet to just fit *inside* the tube.. check and trim to fit nicely inside the tube. Wipe it with alcohol to clean it and hold around the edge as you drop it into place into the tube flush and flat against the end of the mirrors.

Use the Loctite Handyman's glue sparingly and carefully at the edge, do a few spots to get it locked, and allow to dry. Then seal it around the edge, do not miss any spot. Try to keep the area between the two mirrors nice and clean.

Next, gather transparent, but colored objects to use for the pattern bits... nip off bits of vitamin bottles in transparent colors, clear type plastic notebook dividers, glass bits, anything colored but transparent may show some colors through. Bits of non-transparent material will only look dark, and the color won't show.

Get the piece of milky plastic and cut a circular piece that will fit over the end of your tube. This will be the lower cap for the Kaleidoscope.

Temporarily put some of your pieces into the end chamber, try to mix some transparent colors, so the colors can show. Probably no more than 40 or 50 small specks is about right. Place your milky plastic on temporarily and view towards a light ceiling or some light source, see if it looks adequate for a nice display view of the pattern.

Add or subtract a few particles if needed. Once it looks good, apply some clear adhesive sparingly to the rim of the tube, and glue on your milky plastic piece. Keep the Kaleidoscope upside down to dry, so no particles will get trapped in the glue.

There are certain sized vitamin bottle caps which may fit onto your tube to dress it up... all you must do is cut out the center using your Coping Saw or the Hole Saw... if using a Coping Saw, drill a hole in the cap, and take the saw apart. Insert the blade through the hole and re-assemble the saw.

Fig. 90. Lower Kaleidoscope Cover

Cut out the center of the cap, leaving only a bit of the rim. Once the cap is finished, it forms a very nice rim for the lower end of the Kaleidoscope.

If using a Hole Saw, pick a size which leaves a narrow rim of the top cap. See Figure 91.

Fig. 91. Bottle Cap Rim Installed

The last thing to finish the Kaleidoscope is to put on a decorative paper over the tube, and glue on the plastic bottle rim. Paint the top eyepiece with a wide permanent marker. This finishes the Kaleidoscope. See Figure 92.

Fig. 92. Finished Kaleidoscope

Chapter 9

Making a Wooden Toy Train

Parental Help Advised

Items needed: Hardwood Pieces, 1" dowel,
Wheels, 1" diam. from hobby store,
Axle Pegs for 1" wheels,
Small metal screw eyes

This chapter covers making several little wooden train cars, also with a locomotive – using hardwood blocks. Always a great toy for young children.

****

Note: At this point something should be pointed out to parents and kids ... these pieces are all 1 1/2" wide, and almost every piece is square cut, so a Table Saw would be excellent for cutting them. However this is an EXTREMELY dangerous saw... it is NOT for children or an amateur to use. Only an adult person who is practiced in using such a saw should cut with it. Dad or Mom ONLY!!

Even though I had been using a Table Saw for 40 years, I recently got a cut that took 6 stitches inside and 9 outside, and

it went in 1/2" deep between the thumb and first finger. This is a minor cut on a Table Saw, <u>usually a thumb is totally gone</u>, entirely cut off! Think about that.

BE SAFE. Do not use dangerous tools … Get parents to help you, and ALWAYS wear safety glasses--- and ear protectors and dust mask if needed.

****

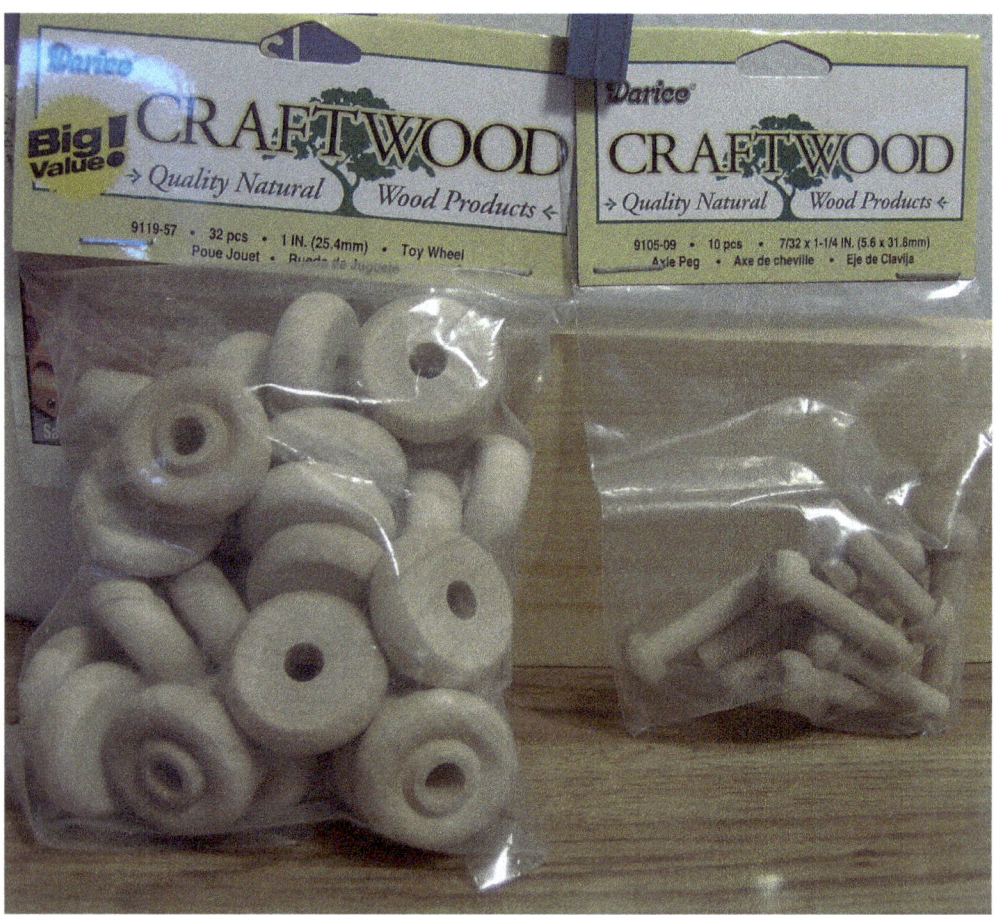

Fig. 93. One Inch Wheels and Axles

Figure 93 shows the wheels and pegs used for these train cars.

For this project we will use small turned wheels and axle pegs available from a craft store. There are many sizes and varieties available but in order to make a smaller size train, these are ideal.

Using hardwood blocks of maple, Oak or Ash with the peg type axles will give strong construction; this train will last a lifetime and even if a part is broken, it can be repaired.

These are very well made wheels, turned and precision, probably made of Birch, which is a very nice hardwood.

Several car types will be shown and it is your choice as to which you build:

- Locomotive

- Caboose

- Tender

- Boxcars

- Flatcars

- Tank Cars

- Log Carrier

Basically in building these cars, you will use a strip of wood 1 1/2" wide, some will be cut from 1/2" thick material... this is common as drawer sides material... and other pieces will be of 1 1/2" stock.

The dimensions here are for a small train using turned 1" diameter wheels, but if you wished a different size block size for the cars, anything that is close will look fine. Or you could go to totally different wheels.

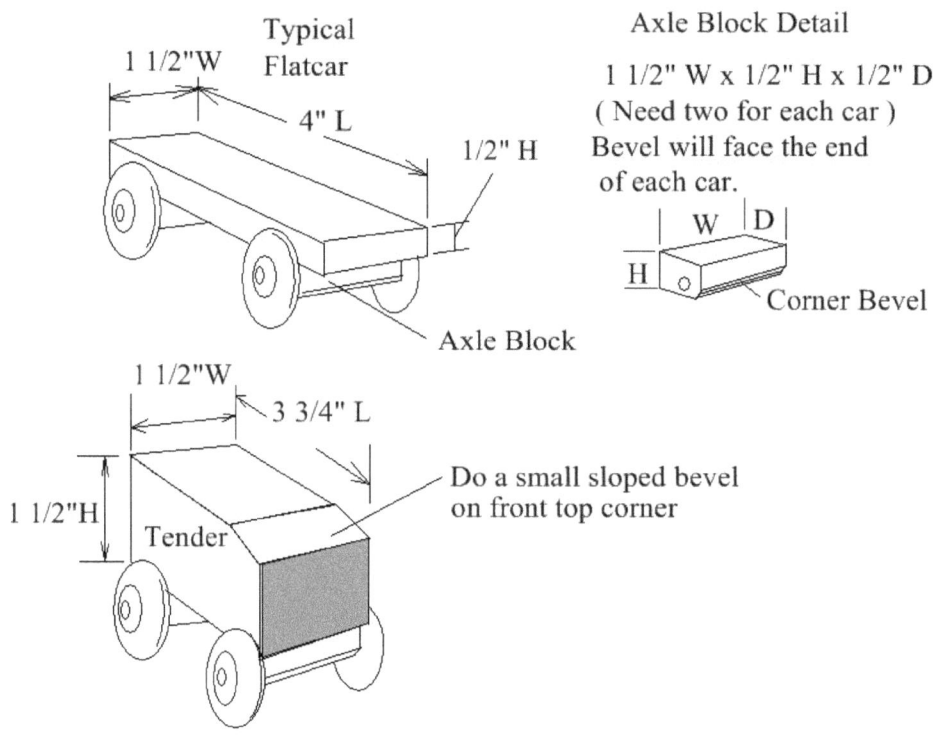

Fig. 94. Simple Car Details

I did these cars about 1 1/2" wide and about 4" long. Anything close will be fine. I did use Ash blocks for the axle blocks, but some of the cars themselves are of Fir. Whatever you have around is fine. It is a wonderful way to get rid of scrap pieces if you are a woodworker!

Basic Train Cars

- 1 1/2" thick wood is great for boxcars, the Tender and the Caboose... perhaps split from a 2 x 4 or purchased as

2 x 2. All you must cut is the length except on the Tender.

Fig. 95. Attaching Axle Blocks

- 1/2" thick pieces are great for the flatcar type cars, and you can add a large dowel to make a tank car, or small dowels to make a car with "logs."

- Width will still be 1 1/2".

See Figure 94 to see some basic ideas. The small axle blocks are meant to glue on the bottom of each main block, about 1/8" from each end. The rounded beveled edge should be at the very bottom and facing the ends. The bevel is there to be certain a corner doesn't protrude past the wheel. Sanding a bit on the corner edge with a Belt Sander will round and bevel it nicely.

Then apply glue and clamp as in Figure 95.

Locomotive

The Locomotive begins as a simple flatcar. By gluing a cabin block on the rear and a large dowel on the front, you create a basic locomotive look. Probably a 1 1/4" diameter dowel is a close size for the boiler. Make it about 1 1/2" less than your total car length. Before you glue the dowel on for the boiler, bevel the front of the flatcar piece sort of like a cowcatcher, which should leave about one inch for the cabin.

You can do a vertical 3/8" or 1/4" dowel stub into the top center of the boiler at the front for a smokestack. Rather than give dimensions here, use your own preference to decide on your dimensions. See Figure 96 for the one done for this book. I also used Danish Oil for this toy, producing a nice look to the train cars. Using the nice, turned wheels really adds to the look, much like the classic wooden trains of decades ago. Playskool or others may even still make them.

Next is a photo of a flatcar set up with small dowels as "logs", also one made into a tank car by adding a large dowel, with a partially cut flat on one side as the glue surface.

As you look at the next group of photos, you will gain confidence in doing your own versions.

Fig. 96. The Locomotive

Fig. 97. Log Carrying Car

Fig. 98. Tank Car

Fig. 99. Stubby Tender

Fig. 100. Boxcar

Fig. 101. Caboose

Fig. 102. Lumber Car

Once your cars are finished, you should use some Danish Oil or similar to finish them; sand nicely and apply according to the can.

When mounting the screw eyes on your cars, follow this guideline:

- Mount a small screw eye on one end of each car except the Locomotive.

- Try to mount the eyes the same distance above the wheels; set midway vertically on the 1/2" base piece on any flatcar style car.

- Then on cars which are 1 1/2" tall, find the same level to match to a flatcar and mount that eyelet or hook.

- On soft woods, simply make an indent with an Awl and screw the little eyelet straight in.

- On a hardwood car, you may drill a tiny bit in with a drill to ease the eyelet mounting.

- Take two small pairs of pliers and carefully bend some of the screw eyes open to form a small hook. Mount one hook on the rear of the locomotive, and one on each car except the caboose, opposite the eyelet end.

This completes the cars so you may connect them together. The final train should look like Figure 103 .

Fig. 103. Entire Train

Chapter 10

A Real Barometer

Items Needed: Wide mouth glass jar,
Toothpick or thin stick,
Balloon,
Rubber Band,
Piece of White cardboard,
Wood scraps

    A Barometer is a scientific device to help forecast local weather. Of course the U.S. Weather Bureau has extremely fine instruments in their facilities, to give them advance warning of weather changes.

    It turns out that if a storm is coming, the air pressure drops, (Barometric Pressure...) and if fair weather is coming, the air pressure increases. This is the principle that is applied for using a Barometer as a weather instrument.

    As a kid, I saw an article on building a homemade Barometer ... it was basically an inverted bottle with a stopper and piece of tubing running up the side, partially filled with colored water. Essentially the two levels -in the bottle and in the tube--were equal when first filled. Then the tube level would rise or fall depending on the Barometric Pressure, thus giving an indication of coming weather.

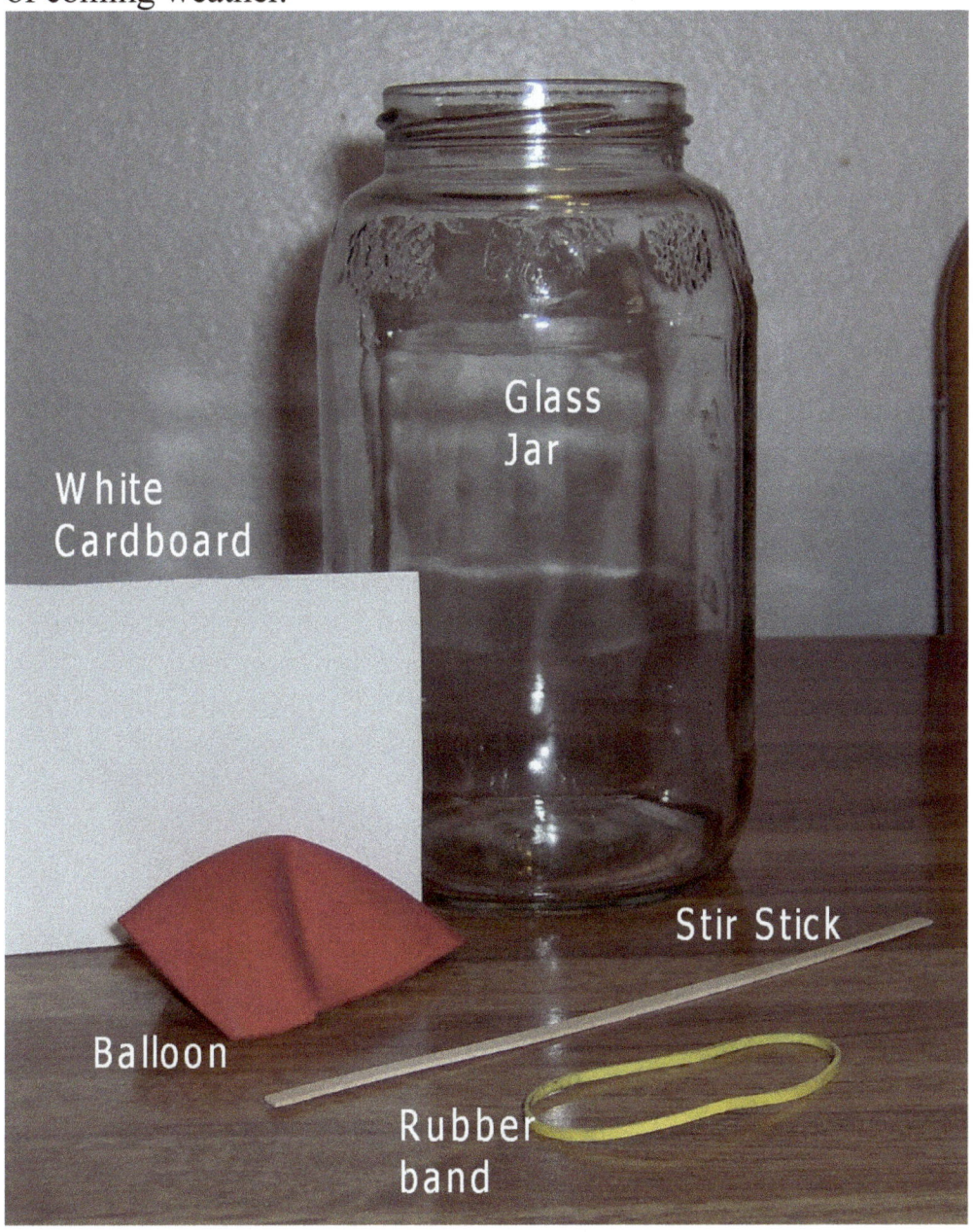

Fig. 104. Parts For Barometer

The Barometer described here is very simple: it consists of a glass bottle with a wide mouth; a balloon is stretched over the top and held with a rubber band. A toothpick is glued to the balloon as a pointer with the white cardboard making up a scale.

The principle is the same. The balloon is stretched over the top of the jar, say, on a reasonably nice day... and locked with the rubber band. The pressure inside and outside the jar is equal.

The stick is glued lightly onto the balloon with its one end in the center of the balloon. The rest of the stick stands out horizontally with the scale made up as shown.

You can do a very nice job by building a base and vertical panel to attach the dial. Glue the vertical panel alongside the bottle. The vertical panel is attached to the base and the base can incorporate blocks to hold the jar in position.

If Barometric Pressure drops, (Storm coming...) the balloon bulges upward, and the pointer indicates low pressure.

If pressure rises, (Fair weather...) the balloon sinks somewhat into the bottle, and the pointer indicates high pressure.

NOTE: You should keep your Barometer in a fairly constant temperature area of your house as varying temperature could also affect the reading. The glass bottle is required rather than a plastic bottle as it is much more stable with pressure or temperature variations.

The balloon over a period of time would equalize due to tiny virtually microscopic holes. But it is perfect for the typical short term pressure changes as storms pass through.

Figure 106 shows the complete Barometer... a good start towards a weather station.

Fig. 105. Basic Barometer

Fig. 106. Base Assembly

If you wish to secure it and make it a bit nicer, see Figure 106. By using a scrap wood piece for a base and mounting a vertical panel for the dial, you can make a nice assembly. A wooden hoop of stir sticks and some blocks of wood to stabilize the jar makes it a complete unit.

Chapter 11

A Rubber Band Gun

<u>Items needed:</u> Piece of 3/4" thick wood,
Wooden Clothespin,
1/4" hardwood, 1/4" wood panel,
Baling Wire,
#4 Box Nail,
7/16 " dowel

In the early 1950's, I lived in Bremerton, Washington. A favorite pastime amongst us kids at that time was playing "Kick the Can", and chasing around with rubber band guns. At that time we used loops of rubber from inner tubes. A clothespin to grab the loop was mounted on top, sometimes on the sides for more shots. We spent many hours playing with those fun toys!

This chapter shows the construction of a simple rubber band gun, either a rifle or pistol style. A lot depends on what size bands you get... all office supply stores carry them. You also can tie multiple bands together if you wish for a rifle style. The principle is the same.

The gun will have a working trigger setup so it is kind of neat for that reason-- more realistic and fun.

As in older styles of rubber band guns, this one uses a clothespin on top. The drawing opposite is not very clear, but later photos will clear everything up.

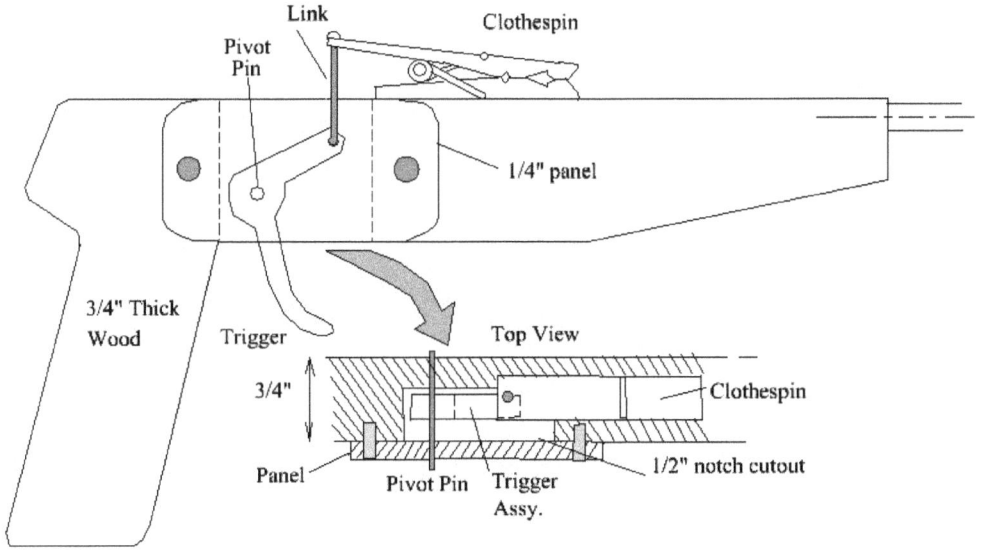

Fig. 107. Basic Gun Principle

The basic material to make the gun is 3/4" wood, probably 1 x 6 standard lumber would be about right. You can do a rifle sort of stock or a pistol. Both will use the same trigger method.

First you must cut out the basic outline of the gun, with the Sabre Saw. If a block piece is cut out of the side of the gun shape similar to the drawing, it gives a space for an inverted "L" shaped trigger assembly. This portion can be cut with the Coping Saw. It is called the receiver.

The lower arm of the trigger is curved to be comfortable with your finger. The upper part of the lever portion will be tied to the clothespin upper end, so as the trigger is pulled, the right arm of the assembly rotates clockwise and opens the clothespin to release the rubber band. The trigger piece should be cut from a sturdy wood with the Coping Saw.

If you do a rifle type stock, you can use multiple rubber bands tied together. The gun described here will be a pistol convertible to a rifle version so the photos will show that double version.

The Pistol First

Fig. 108. Cutting out the Receiver

Draw a version of the pistol receiver in the previous sketch onto a 1 x 6 piece of board. But suit the pistol grip handle to your preference if you wish. *You should leave the rear portion behind the pistol grip uncut if you are planning on a rifle version*; but cut the front portion out using the Sabre Saw or Coping Saw.

Fig. 109. Cutting Halfway for Trigger Assembly

Next, cut partway into the side trigger area, about ½ way deep into the 3/4" width with the Coping Saw . Make the two cuts, one at each end of the area that will be removed, then add one more cut close to one of your cuts and between the two previous cuts. This cut forms an area that will be cleaned out to allow a space to turn the saw sideways and remove the area encompassed by the two outside cuts.

Look at the photos on the next few pages to see the steps. You can also use a drill bit as shown to add holes and weaken the portion to be removed.

Fig. 110. Drilling Across For Easier Sawing

Fig. 111 opposite, Cutting Across the Area

Fig. 112. Trigger Area Removed

Fig. 113 opposite: Cleaning With Chisel

Fig. 114. Sanding the Trigger Area

Once you have cleaned out the trigger area, it is time to make the trigger. Use a good piece of a hard wood, like Maple, Beech or Ash in a thinner thickness than the depth of the slot area you have cut in the work thus far.

Use the Coping Saw (for good control..) to cut out a trigger similar to Figure 115. Clamp the longer piece in a vise and carefully cut the shape.

You must size the trigger to your actual gun, but it will probably be quite close to the one in the figure. About 1/4" thick is a good thickness of this piece. Orient the wood so the lower finger grip portion of the trigger is basically aligned along the wood grain. You can make the trigger a bit thicker too. Be sure the lever arm portion of the trigger pointing right in Figure 115 is sufficient to have a 3/32" hole drilled into it with enough strength remaining to run a wire through it and operate the clothespin. You will see this in a later picture.

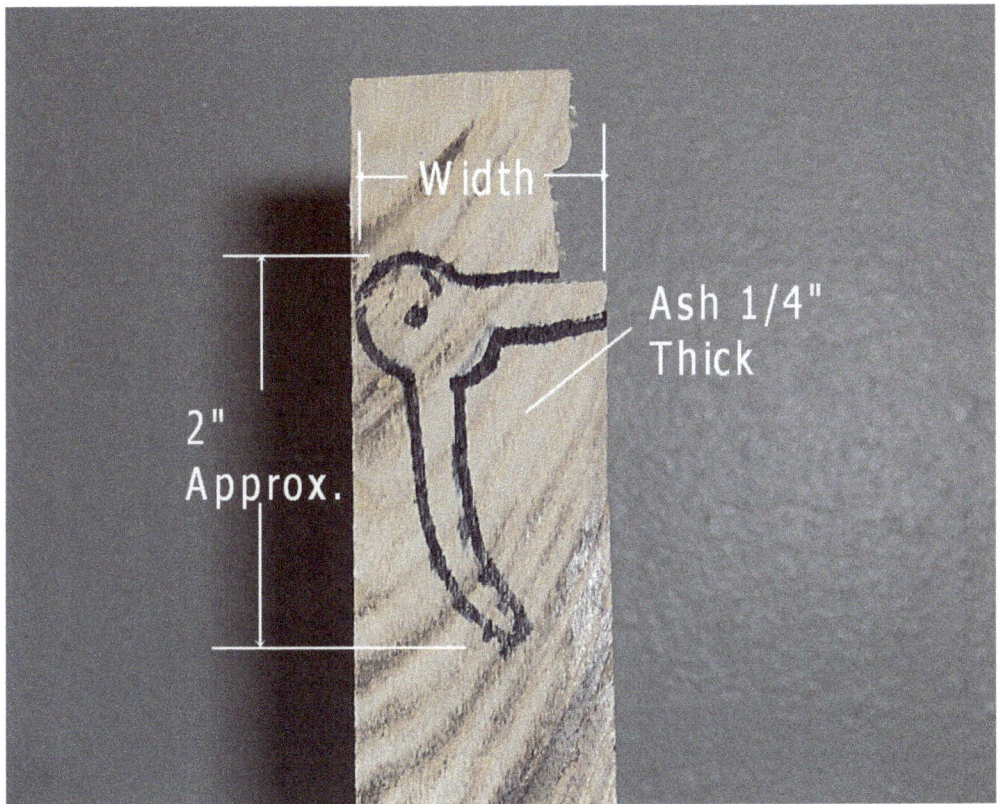

Fig. 115. Typical Trigger

Next get a nice wooden clothespin, and cut about 1/2" from what will be the lower pinch portion. Glue it to the receiver portion as shown in Figure 107 so the cutoff edge is even with the cutout. You may dowel the clothespin and receiver together with a small 1/8" dowel for extra strength if you wish.

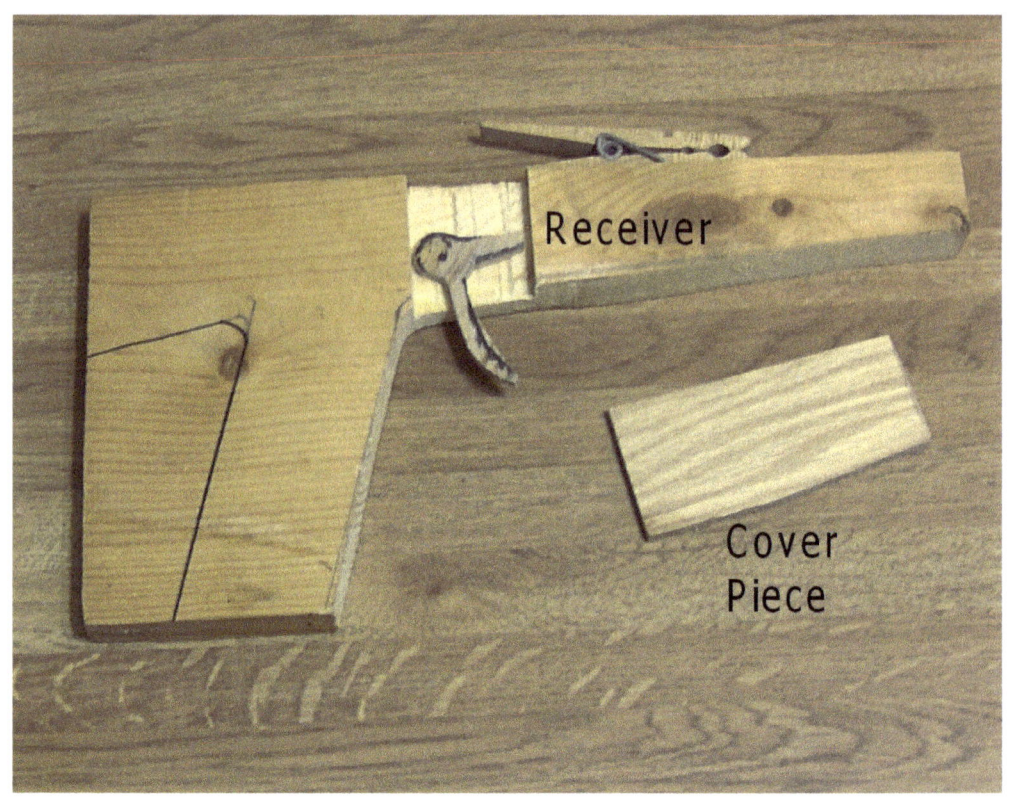

Fig. 116. General View of Receiver

At this point, the parts in Figure 116 are mostly made. By cutting out the remaining rear of the handle you would have the basic body of the pistol.

Cut out a cover piece of 1/4" plywood for the trigger area ... this can be drilled so as to screw to the receiver portion of the pistol.

Drill the holes initially with a 3/32" bit as you hold the pieces aligned in a vise, then enlarge the cover holes to 9/64" and countersink them using a Countersink to bevel the holes for the 3/4" Flathead Screws-- see Figure 117. Do an initial mounting of your cover to be sure of the fit. Then remove and set your trigger into position. You want the trigger hanging below for a

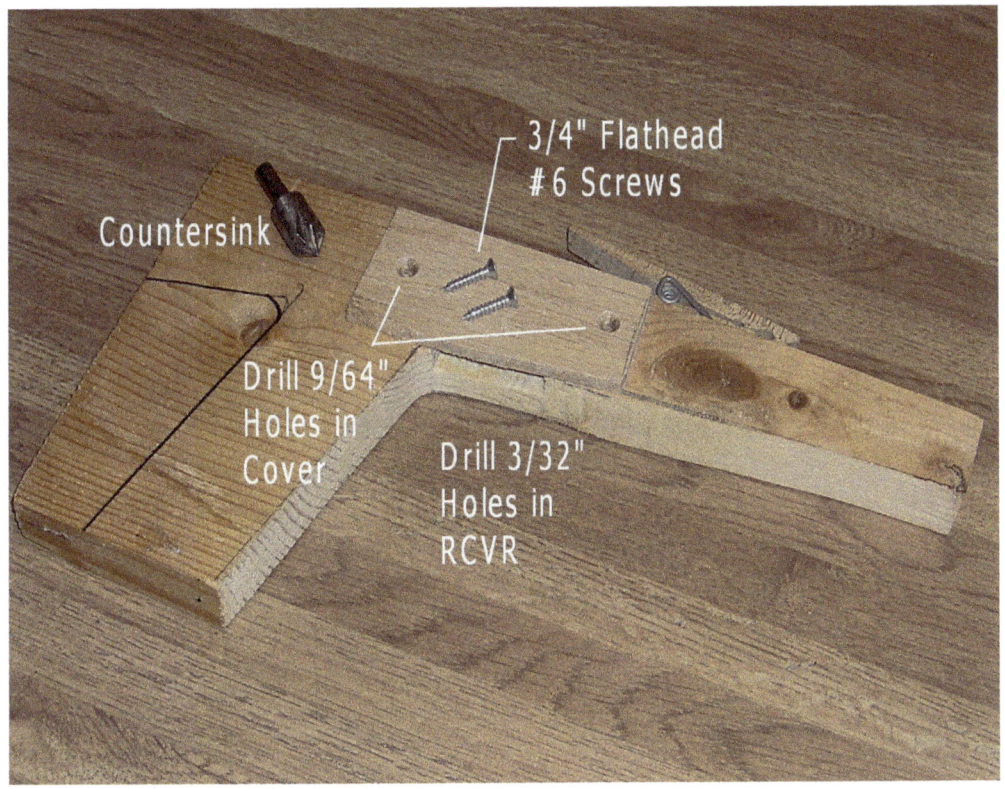

Fig. 117. Cover Details

comfortable feel. You want both the rear and the front lever arm to clear the cutout area.

A good point would be about 1/3 up from the bottom. Drill a pivot hole through the trigger with a 3/32" bit, then set it in place on the receiver and keeping your clearance front and back, mark the receiver notch with the Awl for drilling. Drill as straight as possible through the receiver **with a 3/32" bit.**

Mount the cover plate, turn the pistol over and drill on through the back hole just done through the cover plate keeping the drill as straight as possible perpendicular to the pistol surface.

All we need now is the link setup tying the clothespin to the trigger! A short length of Baling Wire is perfect for the link.

Drill a 3/32" hole about 1/4" from the tip of the upper clothespin arm. Do the same down through the end of the trigger arm.

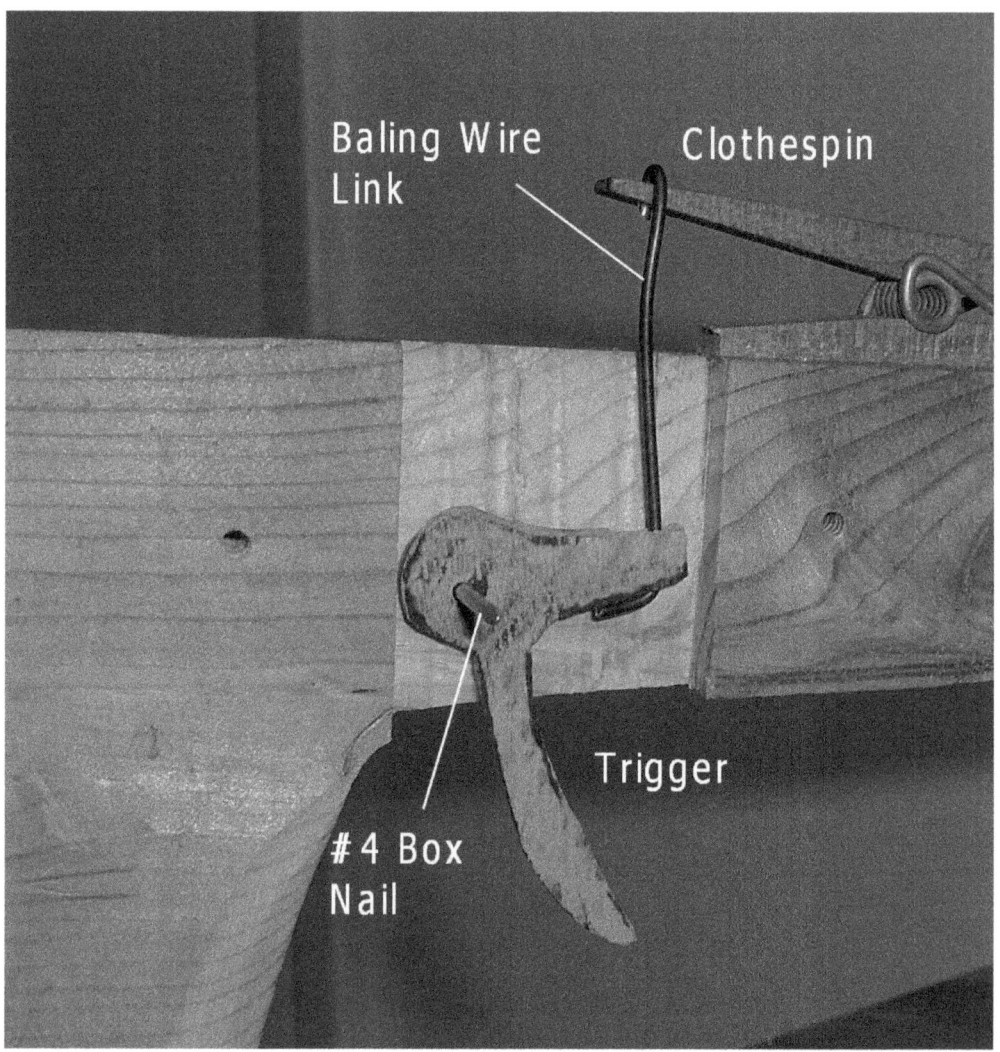

Fig. 118. Detail of Trigger Setup

Figure 118 shows the detail of this operation with the Baling Wire link installed. As you pull the trigger, the link operates the clothespin to release the rubber band.

Note how the link is positioned to the right side of the clothespin so it goes nicely into the trigger space without binding. Bend the Baling Wire to match the underneath part of your trigger, push the wire link through it and with it in the right position, determine the length of the link to curve over and grab the clothespin. Bend your "fish hook" to grab the clothespin and nip off the excess wire. Connect as in Figure 118.

If you wish, you can glue a small scrap above the right facing trigger link arm to prevent reverse rotation loosing the link connection to the clothespin.

Depending on the size of your rubber bands, you can now add a "barrel" on the front of your gun to allow stretching of the bands. 7/16" or 1/2" dowel would be fine. Drill a suitable hole straight into the end of the pistol close to the top and glue in your desired length of barrel. Notch the end a bit so the rubber band will grab into the notch.

If you have not yet cut off the rear of the gun to finish the handle, do that now. Bevel the pistol grip so that it is rounded on all corners to feel rounded in your hand.

Hold the cover over the trigger mechanism and guide your # 4 nail from the rear up through the trigger and through the cover.
****

Caution

You can then make a little file mark to set the length, remove the nail and cut off the tip. Have Dad help you here... ***Protect your eyes as you do this, nippers will shoot the nail out like a bullet!*** File the end of the nail slightly to remove the burr.
****

Reassemble the trigger setup with the # 4 nail coming up through and apply a bit of glue on the back side to glue the nail onto the receiver back side. Screw on the receiver cover once the

glue is dry.
Figure 119 shows the pistol now.

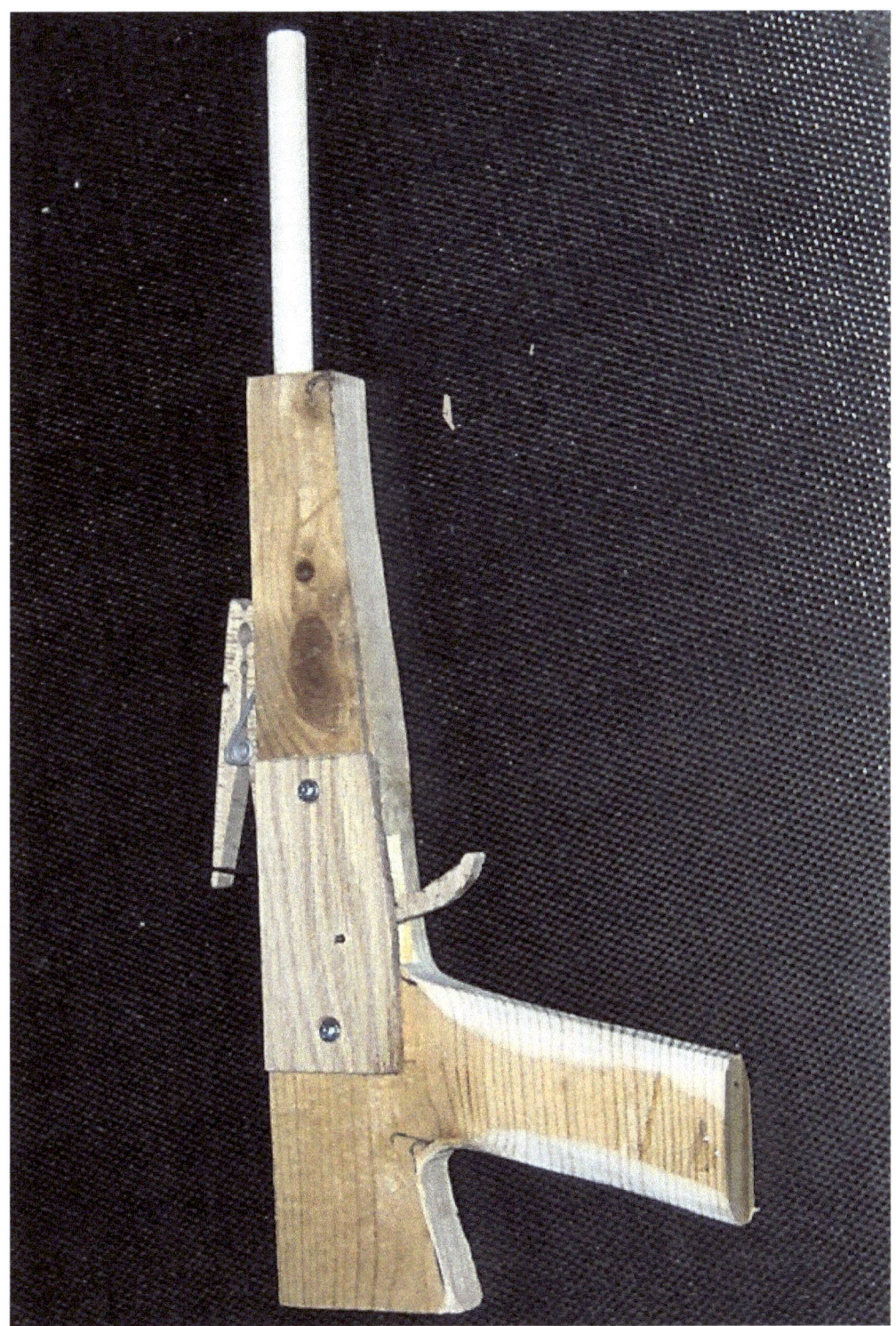

Building The Rifle

The rifle is basically the pistol with a slightly longer barrel if desired and a skeleton lightweight stock. The shoulder pad is made of 3/4" pine or similar and is mounted to the gun receiver by two 7/16" dowels. The dowels are strong and only about 8 1/2" long. See Figure 120.

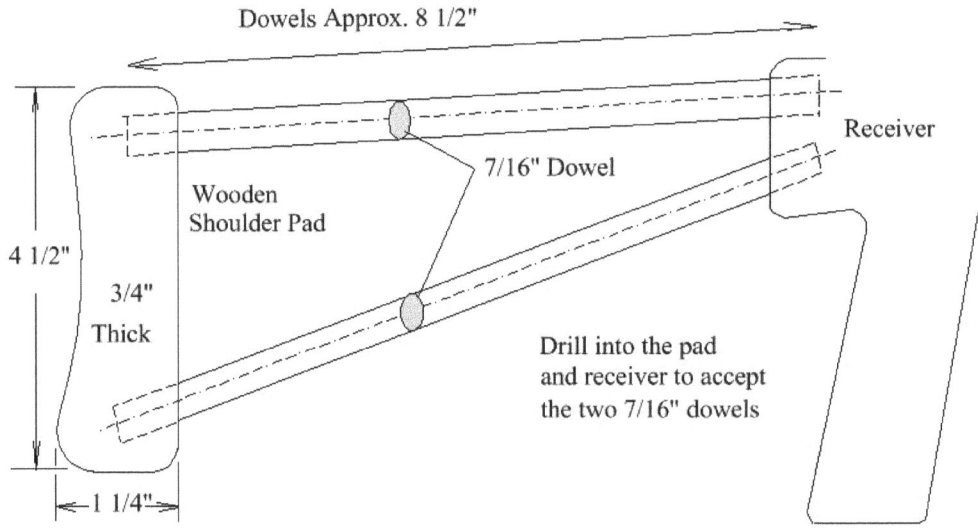

Fig. 120. Stock for Rifle

Bevel and round the edges of the pad, using the Belt Sander. Cut and glue the dowels into the pad after drilling holes as shown. Also drill the receiver portion of the pistol already made, and glue those dowels in as well to make the entire rifle.

If you go to a lumber yard area or construction site you probably can find a small piece of steel banding that was used to band lumber together. Or check with a hardware store... many heavy products are banded with this type of steel banding. It is a thin steel strapping about 1/2" or so in width. It is strong and ideal for a trigger guard for our rifle.

Fig. 121. Metal Pallet Strap

Center punch and drill a 1/8" hole about 1/4" from one end. Bend a tab at the end to mount to your pistol grip. Check the length to bend your strip into a trigger guard. Cut with Tin Snips, and drill the other end. Drill a 3/16" hole into each point on your rifle to mount the trigger guard. File the ends carefully to remove any sharp burrs, and attach it.
The rifle is shown on the next page... Figure 122. The finished version shown is 22 1/2" long and 5 1/2" high at the pistol grip. The barrel sticks out 4".

Chapter 11

Whimmydiddle, Gee-Haw Propeller Stick

Items needed: 7/16" diameter dowel or 10" long stick, about 5/8" x 5/8" square on the end, # 4 Box Nail and wood piece for a propeller, About 10" long 1/4" diam. dowel

This toy is an amazing gadget, besides being a historic artifact of many possible origins. If you can, research it on the internet; the gismo has a long history and apparently no solid background as to its true origin.

It consists of one stick with small notches along the length, also a small propeller on the end. Rubbing the notched portion with another stick makes the propeller spin like crazy, and on uttering a "magic" word, the propeller instantly reverses!

The toy has been attributed as originating from Appalachian mountain people, but who knows as to a possible origin even before that! If you look at some of the pictures of these toys on the internet, you will see some are made from rough sticks, certainly not trimmed lumber.

An internet article on Wikipedia states it is important to have the propeller stick unbalanced in weight, however I believe the most critical item is that the hole in the propeller must be large compared to the nail it pivots about.

Fig. 123. Whimmydiddle Toy

A very balanced propeller works just fine, and actually meets the slight imbalance criteria claimed if it operates on a small nail compared to the propeller hole. The main stick should be made first. It can be a dowel about 7/16" in diameter or a long stick with a cross section of about 5/8" square at the end. It should be about 10 to 11" long.

With a ruler, mark the stick or dowel at 1/8" intervals for about a 6" length in the center portion. Leave a gap at the ends as in Figure 123.

Using either the pocket knife or the Coping Saw, carefully make small "V" notches at each 1/8" mark as you move along the stick cutting them carefully.

Once that is done, slightly taper the propeller end a bit for looks. Drill a 1/16" hole about 1" deep straight into the propeller end of your notched stick. Later when you place the # 4 nail, the stick will not split.

Cut a small 2 1/2" long 1/2" x 1/2" cross section stick for the propeller. You can shape it like a propeller or not, your choice. The one in Figure 123 was cut like a propeller leaving a 1/2" uncut portion in the middle as a nice area to drill.

Drill the propeller hole large, about 7 /64" or 1/8" or so right in the center of your propeller piece. Trim the propeller a bit so it balances fairly well when the # 4 nail is placed through the center.

Gently apply the # 4 Box Nail through the propeller and tap into the tip of the notched stick, leaving it free to spin. Make sure it spins freely.

The next part takes some practice but is fairly easily mastered. It is learning how to make the propeller spin--- and reverse instantly.

Basically you first take hold at the end of the notched stick with one hand opposite the propeller. Take the 1/4" dowel in your other hand with about 6" of the main length of the dowel facing out forward with your thumb bracing against it. Hook your first finger behind the dowel and spaced out a bit from your thumb. Look at Figure 124.

You must learn to hold your first finger as shown, and spaced a tiny bit further from your thumb than the width of your notched propeller stick.

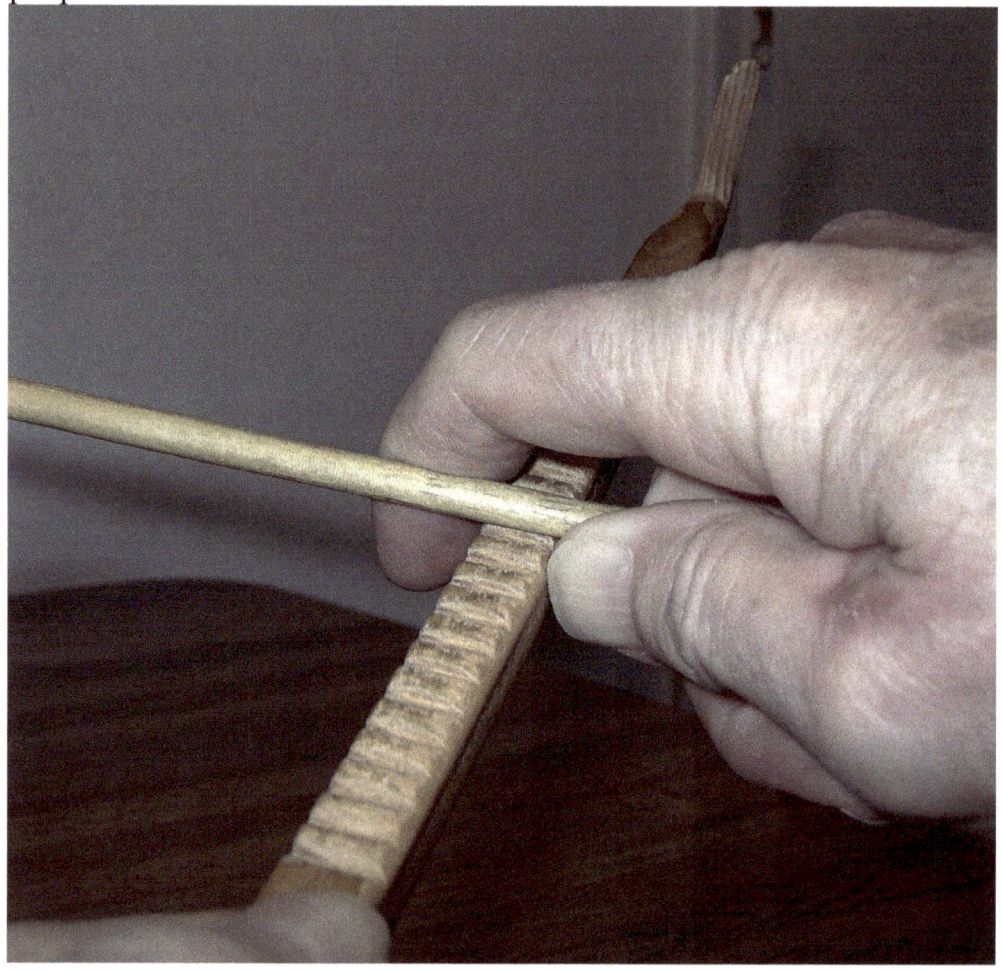

Fig. 124. Grip of Operating Stick

Look at Figure 124. Here is how you make it operate; practice slowly until you learn the technique. Hold the propeller stick at the tail end with the one hand:

- As you gently rub the dowel back and forth on the notched

propeller stick with your other hand, pull on the side of the propeller stick with the tip of your first finger that is looped over the top. Make sure to keep your thumb clear from touching; be sure only your first finger pulls and rubs on the side as you operate the dowel stick back and forth... The propeller will spin nicely!

- To make the propeller reverse, slide your hand slightly crosswise on the propeller stick as you are rubbing the dowel back and forth, so that now your thumbnail pushes against the propeller stick and your finger **does not**. Now the propeller reverses and goes the other way!

Practice operating the stick until you have it down. A nice thing to show your younger brothers and sisters as a "Magic Propeller."

(Evidently the sometimes name of Gee-Haw stick had to do with the commands given to horse teams in the old days to go left or right. A good name!!)

Chapter 12

Final Notes

These projects should give you a fun start to making your own toys. There are so many things you can do without ever having to buy a toy from a store.

If you have tackled some of these toys, you will know your friends are impressed and maybe interested in trying to build them too. Help them. The fun of building your own toy is unmatched by anything else, knowing you can build a great toy YOURSELF.

The **use of tools safely** will guide you as you become an adult and will give you a hobby towards making thing as you become a Dad. And you can pass that along to your kids.

As a boy, I built many of these toys for myself growing up and then later as a Dad, I built many of these sorts of toys for my two sons, and now Grandkids.

Toys not shown here were an Army Tank which had some springs to fire a wooden peg when cocked and released, and two Bulldozers which had wooden segmented tracks. As you gain confidence, you can learn and try new gadgets in your toymaking, too.

If you see a picture of a toy you like, you can judge its proportions and make a good copy. Here is an example: Look at the picture on the next page.

It is obviously a tank of some type; I did it for my previous toymaking book.

Suppose you wished to make this tank. If it were in a toy catalog, the dimensions might be given, so you could start there.

Eyeball or Ruler Ratios

This tank has wheels 1 5/16" approximately, a tweak smaller that the dog toys in Chapter 5. Remember, those wheels were 1 3/8" diameter. Knowing that, you can use a ruler or eyeball and compare the diameter of the wheels to the thickness of the tank body.

In the picture, you find that the board making up the tank body is about 2/3 the wheel diameter. And a common board thickness is 3/4" very close to that as 2/3 of 1 5/16 is almost exactly 3/4" ! Exactly correct.

Next, compare the turret to the wheel size... your eyes are very good at seeing the ratio even though the angles are different ... if you guess the diameter ratio based on a line front to back of the tank (to give a more accurate comparison..) for the turret versus the wheels, doesn't it look about 1 ½ times the wheel diameter? That comes out about 2" which is indeed the correct value.

Looking lengthwise again, can you see the tank body length is almost 3 times the diameter of the turret? Guess what, the tank body is close, about 5 1/2". (A standard 1 x6 width. Note the grain agrees to a cut off board 5 1/2" wide...)

Finally for the width, use your own feeling of a typical tank rectangular shape. Draw two lines on paper to show your 5 1/2 " to 6" length and decide on a pleasing width ... chances are you will be about 3 1/2" on width.

The wheels are thick, 3/4" actually. All you have to do now is cut out these pieces, then you can place them in position and decide on the axle blocks to raise the wheels a tad away from rubbing the body of the tank. Use 1/4" axles of course and the holes in the axle blocks should be a hair under 5/16".

Your blocks can be made of 3/4" thick wood, a bit narrower than the width of the tank, and when glued on, should allow say 1/4" clearance so the wheels won't rub on the body above. Length should be about 2" to allow the wheels to sit beneath the body a bit.

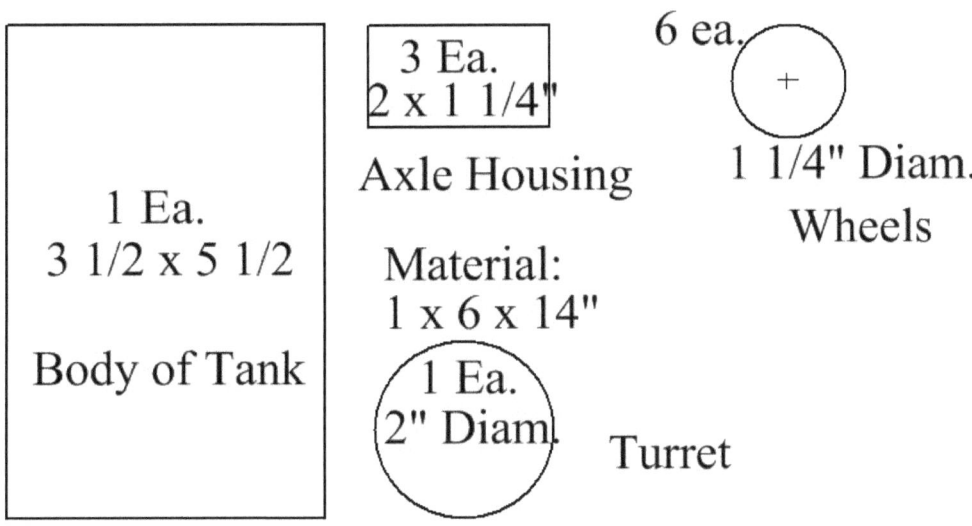

Fig. 126. Actual Pattern from my 2008 Book

This method works for toys you see in pictures, give it a try and you will be doing your math at the same time !

And above all be safe... use the safety items, safety glasses, dust mask and ear protection as necessary.

Have fun!!

www.ingramcontent.com/pod-product-compliance
Lightning Source LLC
Chambersburg PA
CBHW040730020526
44112CB00058B/2911